Praise for THE Joy *of* x

"I loved this beautiful book from the first page. With his unique ingenuity and affable charm, Strogatz disassembles mathematics as a subject, both feared and revered, and reassembles it as a world, both accessible and magical. *The Joy of x* is, well, a joy."
—Janna Levin, professor of physics and astronomy, Barnard College, and author of *How the Universe Got Its Spots* and *A Madman Dreams of Turing Machines*

"*The Joy of x* literally breezes through every one of these scary-sounding math subjects and makes it quite digestible and fun. And it's entertaining reading, covering how math relates to even zebra stripes and sunsets and dragonfly wings, even your dating life."
—NPR's *Science Friday*

"Strogatz may be the only person alive with the skill to Pied Piper me into the murky abyss of set theory. I literally learned something on every page, despite my innumerate brain. This is a fantastic book, conveyed with clarity, technical mastery, and infectious joy."
—**Jad Abumrad, host of** *Radiolab*

"A delightful antidote to the math phobia that infects most students exposed to the standard curriculum . . . *The Joy of x* presents the essential ideas of the major branches of math in engaging and entertaining language."
—*Science News*

"Amazingly, mathematicians can see patterns in the universe that the rest of us are usually blind to. With clarity and dry wit, *The Joy of x* opens a window onto this hidden world with its landscapes of beauty and wonder."
—**Alan Alda**

"This book is, simply put, fantastic. It introduces the reader to the underlying concepts of mathematics—presenting reasons for its unfamiliar language and explaining conceptual frameworks that do in fact make understanding complex problems easier. In a world where mathematics is essential, but largely poorly understood, Steve Strogatz's teaching skills and deft writing style are an important contribution."
—**Lisa Randall, professor of physics, Harvard University, and author of** *Warped Passages* **and** *Knocking on Heaven's Door*

"Strogatz has discovered a magical function that transforms 'math' into 'joy,' page after wonderful page. He takes everything that ever mystified you about math and makes it better than clear—he makes it wondrous, delicious, and amazing."
—Daniel Gilbert, professor of psychology,
Harvard University, and author of *Stumbling on Happiness*

"Strogatz's graceful prose is perfectly pitched for a popular math book: authoritative without being patronizing, friendly without being whimsical, and always clear and accessible. His *x* marks the spot—and hits it."
—Alex Bellos, author of *Here's Looking at Euclid*

"A neat survey of the major fields of math by a professor adept at writing both popularizations and textbooks . . . A great book for the bright and curious, including even kids at grade school level up to college and beyond."
—*Kirkus Reviews*

"Even the most math-phobic readers might forget their dread after just a few pages of Strogatz's latest."
—*Publishers Weekly*

THE Joy *of* x

A Guided Tour of Math, from One to Infinity

STEVEN STROGATZ

An Imprint of HarperCollins*Publishers*
Boston   New York

First Mariner edition 2013
Copyright © 2012 by Steven Strogatz

Mariner
An Imprint of HarperCollins Publishers, registered in the United States of America and/or other jurisdictions.

www.marinerbooks.com

Library of Congress Cataloging-in-Publication Data
Strogatz, Steven H. (Steven Henry)
  The joy of *x* : a guided tour of math, from one to infinity / Steven Strogatz.
      p. cm.
  ISBN 978-0-547-51765-0 (hardback)   ISBN 978-0-544-10585-0 (pbk.)
  1. Mathematics—Popular works. I. Title.
  QA93.S77 2012
  510—dc23   2012017320

Book design by Yasuyo Iguchi
Illustrations by Margaret Nelson

Printed in the United States of America
23 24 25 26 27 LBC 16 15 14 13 12

Chapters 1–3, 5, 7, 8, 11, 12, 16–18, 20, 23, 26, 28, and 30 are adapted, with permission, from pieces originally published in *The New York Times*.

Grateful acknowledgment is made for permission to reprint an excerpt from the following copyrighted work: pp. 201–202, from p. 111 of *The Solitude of Prime Numbers: A Novel* by Paolo Giordano, translated by Shaun Whiteside, copyright © 2008 by Arnoldo Mondadori Editore S.p.A., translation copyright © 2009 by Shaun Whiteside. Used by permission of Pamela Dorman Books, an imprint of Viking Penguin, a division of Penguin Group (USA) Inc.

Credits appear on page 307.

Contents

Preface

I have a friend who gets a tremendous kick out of science, even though he's an artist. Whenever we get together all he wants to do is chat about the latest thing in psychology or quantum mechanics. But when it comes to math, he feels at sea, and it saddens him. The strange symbols keep him out. He says he doesn't even know how to pronounce them.

In fact, his alienation runs a lot deeper. He's not sure what mathematicians do all day, or what they mean when they say a proof is elegant. Sometimes we joke that I should just sit him down and teach him everything, starting with $1 + 1 = 2$ and going as far as we can.

Crazy as it sounds, that's what I'll be trying to do in this book. It's a guided tour through the elements of math, from preschool to grad school, for anyone out there who'd like to have a second chance at the subject—but this time from an adult perspective. It's not intended to be remedial. The goal is to give you a better feeling for what math is all about and why it's so enthralling to those who get it.

We'll discover how Michael Jordan's dunks can help explain the fundamentals of calculus. I'll show you a simple—and mind-blowing—way to understand that staple of geometry, the Pythagorean theorem. We'll try to get to the bottom of some of life's mysteries, big and small: Did O.J. do it? How should you flip your mattress to get the maximum wear out of it? How many people should you date before settling down? And we'll see why some infinities are bigger than others.

Math is everywhere, if you know where to look. We'll spot sine waves in zebra stripes, hear echoes of Euclid in the Declaration of Independence, and recognize signs of negative numbers in the run-up to World War I. And we'll see how our lives today are being touched by new kinds of math, as we search for restaurants online and try to understand — not to mention survive — the frightening swings in the stock market.

By a coincidence that seems only fitting for a book about numbers, this one was born on the day I turned fifty. David Shipley, who was then the editor of the op-ed page for the *New York Times*, had invited me to lunch on the big day (unaware of its semicentennial significance) and asked if I would ever consider writing a series about math for his readers. I loved the thought of sharing the pleasures of math with an audience beyond my inquisitive artist friend.

"The Elements of Math" appeared online in late January 2010 and ran for fifteen weeks. In response, letters and comments poured in from readers of all ages. Many who wrote were students and teachers. Others were curious people who, for whatever reason, had fallen off the track somewhere in their math education but sensed they were missing something worthwhile and wanted to try again. Especially gratifying were the notes I received from parents thanking me for helping them explain math to their kids and, in the process, to themselves. Even my colleagues and fellow math aficionados seemed to enjoy the pieces — when they weren't suggesting improvements (or perhaps especially then!).

All in all, the experience convinced me that there's a profound but little-recognized hunger for math among the general public. Despite everything we hear about math phobia, many people *want* to understand the subject a little better. And once they do, they find it addictive.

The Joy of x is an introduction to math's most compelling and far-reaching ideas. The chapters—some from the original *Times* series—are bite-size and largely independent, so feel free to snack wherever you like. If you want to wade deeper into anything, the notes at the end of the book provide additional details and suggestions for further reading.

For the benefit of readers who prefer a step-by-step approach, I've arranged the material into six main parts, following the lines of the traditional curriculum.

Part 1, "Numbers," begins our journey with kindergarten and grade-school arithmetic, stressing how helpful numbers can be and how uncannily effective they are at describing the world.

Part 2, "Relationships," generalizes from working with numbers to working with *relationships* between numbers. These are the ideas at the heart of algebra. What makes them so crucial is that they provide the first tools for describing how one thing affects another, through cause and effect, supply and demand, dose and response, and so on—the kinds of relationships that make the world complicated and rich.

Part 3, "Shapes," turns from numbers and symbols to shapes and space—the province of geometry and trigonometry. Along with characterizing all things visual, these subjects raise math to new levels of rigor through logic and proof.

In part 4, "Change," we come to calculus, the most penetrating and fruitful branch of math. Calculus made it possible to predict the motions of the planets, the rhythm of the tides, and virtually every other form of continuous change in the universe and ourselves. A supporting theme in this part is the role of infinity. The domestication of infinity was the breakthrough that made calculus work. By harnessing the awesome power of the infinite, calculus could finally solve many long-standing problems that had defied the ancients, and that ultimately led to the scientific revolution and the modern world.

Part 5, "Data," deals with probability, statistics, networks, and data mining, all relatively young subjects inspired by the messy side of life: chance and luck, uncertainty, risk, volatility, randomness, interconnectivity. With the right kinds of math, and the right kinds of data, we'll see how to pull meaning from the maelstrom.

As we near the end of our journey in part 6, "Frontiers," we approach the edge of mathematical knowledge, the borderland between what's known and what remains elusive. The sequence of chapters follows the familiar structure we've used throughout — numbers, relationships, shapes, change, and infinity — but each of these topics is now revisited more deeply, and in its modern incarnation.

I hope that all of the ideas ahead will provide joy — and a good number of Aha! moments. But any journey needs to begin at the beginning, so let's start with the simple, magical act of counting.

Part One **NUMBERS**

From Fish to Infinity

THE BEST INTRODUCTION to numbers I've ever seen—the clearest and funniest explanation of what they are and why we need them—appears in a *Sesame Street* video called *123 Count with Me*. Humphrey, an amiable but dimwitted fellow with pink fur and a green nose, is working the lunch shift at the Furry Arms Hotel when he takes a call from a roomful of penguins. Humphrey listens carefully and then calls out their order to the kitchen: "Fish, fish, fish, fish, fish, fish." This prompts Ernie to enlighten him about the virtues of the number six.

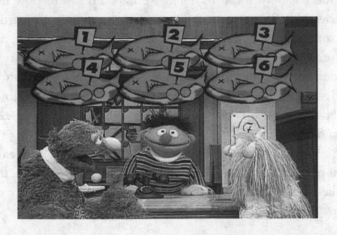

Children learn from this that numbers are wonderful short-cuts. Instead of saying the word "fish" exactly as many times as there are penguins, Humphrey could use the more powerful concept of six.

As adults, however, we might notice a potential downside to numbers. Sure, they are great timesavers, but at a serious cost in abstraction. Six is more ethereal than six fish, precisely because it's more general. It applies to six of anything: six plates, six penguins, six utterances of the word "fish." It's the ineffable thing they all have in common.

Viewed in this light, numbers start to seem a bit mysterious. They apparently exist in some sort of Platonic realm, a level above reality. In that respect they are more like other lofty concepts (e.g., truth and justice), and less like the ordinary objects of daily life. Their philosophical status becomes even murkier upon further reflection. Where exactly do numbers come from? Did humanity invent them? Or discover them?

An additional subtlety is that numbers (and all mathematical ideas, for that matter) have lives of their own. We can't control them. Even though they exist in our minds, once we decide what we mean by them we have no say in how they behave. They obey certain laws and have certain properties, personalities, and ways of combining with one another, and there's nothing we can do about it except watch and try to understand. In that sense they are eerily reminiscent of atoms and stars, the things of this world, which are likewise subject to laws beyond our control . . . except that those things exist outside our heads.

This dual aspect of numbers—as part heaven, part earth —is perhaps their most paradoxical feature, and the feature that makes them so useful. It is what the physicist Eugene Wigner had in mind when he wrote of "the unreasonable effectiveness of mathematics in the natural sciences."

In case it's not clear what I mean about the lives of numbers and their uncontrollable behavior, let's go back to the Furry Arms. Suppose that before Humphrey puts in the penguins' order, he suddenly gets a call on another line from a room occupied by the same number of penguins, all of them also clamoring for fish. After taking both calls, what should Humphrey yell out to the kitchen? If he hasn't learned anything, he could shout "fish" once for each penguin. Or, using his numbers, he could tell the cook he needs six orders of fish for the first room and six more for the second room. But what he really needs is a new concept: addition. Once he's mastered it, he'll proudly say he needs six plus six (or, if he's a showoff, twelve) fish.

The creative process here is the same as the one that gave us numbers in the first place. Just as numbers are a shortcut for counting by ones, addition is a shortcut for counting by any amount. This is how mathematics grows. The right abstraction leads to new insight, and new power.

Before long, even Humphrey might realize he can keep counting forever.

Yet despite this infinite vista, there are always constraints on our creativity. We can decide what we mean by things like 6 and +, but once we do, the results of expressions like 6 + 6 are beyond our control. Logic leaves us no choice. In that sense, math always involves both invention *and* discovery: we invent the concepts but discover their consequences. As we'll see in the coming chapters, in mathematics our freedom lies in the questions we ask—and in how we pursue them—but not in the answers awaiting us.

LIKE ANYTHING ELSE, arithmetic has its serious side and its playful side.

The serious side is what we all learned in school: how to work with columns of numbers, adding them, subtracting them, grinding them through the spreadsheet calculations needed for tax returns and year-end reports. This side of arithmetic is important, practical, and — for many people — joyless.

The playful side of arithmetic is a lot less familiar, unless you were trained in the ways of advanced mathematics. Yet there's nothing inherently advanced about it. It's as natural as a child's curiosity.

In his book *A Mathematician's Lament*, Paul Lockhart advocates an educational approach in which numbers are treated more concretely than usual: he asks us to imagine them as groups of rocks. For example, 6 corresponds to a group of rocks like this:

You probably don't see anything striking here, and that's right —unless we make further demands on numbers, they all look pretty much the same. Our chance to be creative comes in what we ask of them.

For instance, let's focus on groups having between 1 and 10 rocks in them, and ask which of these groups can be rearranged into square patterns. Only two of them can: the group with 4 and the group with 9. And that's because $4 = 2 \times 2$ and $9 = 3 \times 3$; we get these numbers by squaring some other number (actually making a square shape).

A less stringent challenge is to identify groups of rocks that can be neatly organized into a rectangle with exactly two rows that come out even. That's possible as long as there are 2, 4, 6, 8, or 10 rocks; the number has to be even. If we try to coerce any of the other numbers from 1 to 10—the odd numbers—into two rows, they always leave an odd bit sticking out.

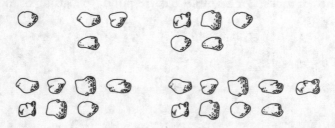

Still, all is not lost for these misfit numbers. If we add two of them together, their protuberances match up and their sum comes out even; Odd + Odd = Even.

If we loosen the rules still further to admit numbers greater than 10 and allow a rectangular pattern to have more than two rows of rocks, some odd numbers display a talent for making these larger rectangles. For example, the number 15 can form a 3 × 5 rectangle:

So 15, although undeniably odd, at least has the consolation of being a composite number — it's composed of three rows of five rocks each. Similarly, every other entry in the multiplication table yields its own rectangular rock group.

Yet some numbers, like 2, 3, 5, and 7, truly are hopeless. None of them can form any sort of rectangle at all, other than a simple line of rocks (a single row). These strangely inflexible numbers are the famous prime numbers.

So we see that numbers have quirks of structure that endow them with personalities. But to see the full range of their behavior, we need to go beyond individual numbers and watch what happens when they interact.

For example, instead of adding just two odd numbers together, suppose we add all the consecutive odd numbers, starting from 1:

$$1 + 3 = 4$$
$$1 + 3 + 5 = 9$$
$$1 + 3 + 5 + 7 = 16$$
$$1 + 3 + 5 + 7 + 9 = 25.$$

The sums above, remarkably, always turn out to be perfect
squares. (We saw 4 and 9 in the square patterns discussed earlier, and $16 = 4 \times 4$, and $25 = 5 \times 5$.) A quick check shows that
this rule keeps working for larger and larger odd numbers; it
apparently holds all the way out to infinity. But what possible
connection could there be between odd numbers, with their
ungainly appendages, and the classically symmetrical numbers
that form squares? By arranging our rocks in the right way, we
can make this surprising link seem obvious—the hallmark of
an elegant proof.

The key is to recognize that odd numbers can make L-
shapes, with their protuberances cast off into the corner.
And when you stack successive L-shapes together, you get a
square!

This style of thinking appears in another recent book, though for altogether different literary reasons. In Yoko Ogawa's charming novel *The Housekeeper and the Professor*, an astute but uneducated young woman with a ten-year-old son is hired to take care of an elderly mathematician who has suffered a traumatic brain injury that leaves him with only eighty minutes of short-term memory. Adrift in the present, and alone in his shabby cottage with nothing but his numbers, the Professor tries to connect with the Housekeeper the only way he knows how: by inquiring about her shoe size or birthday and making mathematical small talk about her statistics. The Professor also takes a special liking to the Housekeeper's son, whom he calls Root, because the flat top of the boy's head reminds him of the square root symbol, $\sqrt{\ }$.

One day the Professor gives Root a little puzzle: Can he find the sum of all the numbers from 1 to 10? After Root carefully adds the numbers and returns with the answer (55), the Professor asks him to find a better way. Can he find the answer *without* adding the numbers? Root kicks the chair and shouts, "That's not fair!"

But little by little the Housekeeper gets drawn into the world of numbers, and she secretly starts exploring the puzzle herself. "I'm not sure why I became so absorbed in a child's math problem with no practical value," she says. "At first, I was conscious of wanting to please the Professor, but gradually that feeling faded and I realized it had become a battle between the problem and me. When I woke in the morning, the equation was waiting:

$$1 + 2 + 3 + \ldots + 9 + 10 = 55$$

and it followed me all through the day, as though it had burned itself into my retina and could not be ignored."

There are several ways to solve the Professor's problem (see how many you can find). The Professor himself gives an argument along the lines we developed above. He interprets the sum from 1 to 10 as a triangle of rocks, with 1 rock in the first row, 2 in the second, and so on, up to 10 rocks in the tenth row:

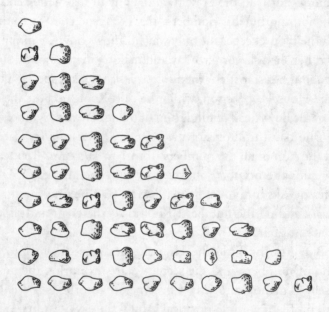

By its very appearance this picture gives a clear sense of negative space. It seems only half complete. And that suggests a creative leap. If you copy the triangle, flip it upside down, and add it as the missing half to what's already there, you get something much simpler: a rectangle with ten rows of 11 rocks each, for a total of 110.

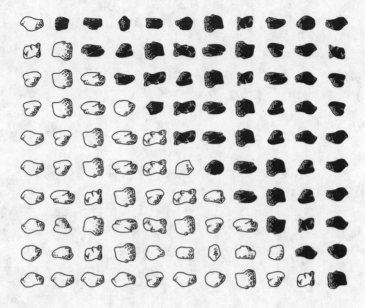

Since the original triangle is half of this rectangle, the desired sum must be half of 110, or 55.

Looking at numbers as groups of rocks may seem unusual, but actually it's as old as math itself. The word "calculate" reflects that legacy—it comes from the Latin word *calculus*, meaning a pebble used for counting. To enjoy working with numbers you don't have to be Einstein (German for "one stone"), but it might help to have rocks in your head.

The Enemy of My Enemy

3

IT'S TRADITIONAL TO teach kids subtraction right after addition. That makes sense—the same facts about numbers get used in both, though in reverse. And the black art of borrowing, so crucial to successful subtraction, is only a little more baroque than that of carrying, its counterpart for addition. If you can cope with calculating $23 + 9$, you'll be ready for $23 - 9$ soon enough.

At a deeper level, however, subtraction raises a much more disturbing issue, one that never arises with addition. Subtraction can generate negative numbers. If I try to take 6 cookies away from you but you have only 2, I can't do it—except in my mind, where you now have negative 4 cookies, whatever that means.

Subtraction forces us to expand our conception of what numbers are. Negative numbers are a lot more abstract than positive numbers—you can't see negative 4 cookies and you certainly can't eat them—but you can think about them, and you *have* to, in all aspects of daily life, from debts and overdrafts to contending with freezing temperatures and parking garages.

Still, many of us haven't quite made peace with negative

numbers. As my colleague Andy Ruina has pointed out, people have concocted all sorts of funny little mental strategies to sidestep the dreaded negative sign. On mutual fund statements, losses (negative numbers) are printed in red or nestled in parentheses with nary a negative sign to be found. The history books tell us that Julius Caesar was born in 100 B.C., not −100. The subterranean levels in a parking garage often have designations like B1 and B2. Temperatures are one of the few exceptions: folks do say, especially here in Ithaca, New York, that it's −5 degrees outside, though even then, many prefer to say 5 below zero. There's something about that negative sign that just looks so unpleasant, so . . . negative.

Perhaps the most unsettling thing is that a negative times a negative is a positive. So let me try to explain the thinking behind that.

How should we define the value of an expression like −1 × 3, where we're multiplying a negative number by a positive number? Well, just as 1 × 3 means 1 + 1 + 1, the natural definition for −1 × 3 is (−1) + (−1) + (−1), which equals −3. This should be obvious in terms of money: if you owe me $1 a week, after three weeks you're $3 in the hole.

From there it's a short hop to see why a negative times a negative should be a positive. Take a look at the following string of equations:

$$-1 \times 3 = -3$$
$$-1 \times 2 = -2$$
$$-1 \times 1 = -1$$
$$-1 \times 0 = 0$$
$$-1 \times -1 = ?$$

Now look at the numbers on the far right and notice their

orderly progression: −3, −2, −1, 0, . . . At each step, we're adding 1 to the number before it. So wouldn't you agree the next number should logically be 1?

That's one argument for why $(-1) \times (-1) = 1$. The appeal of this definition is that it preserves the rules of ordinary arithmetic; what works for positive numbers also works for negative numbers.

But if you're a hard-boiled pragmatist, you may be wondering if these abstractions have any parallels in the real world. Admittedly, life sometimes seems to play by different rules. In conventional morality, two wrongs don't make a right. Likewise, double negatives don't always amount to positives; they can make negatives more intense, as in "I can't get no satisfaction." (Actually, languages can be very tricky in this respect. The eminent linguistic philosopher J. L. Austin of Oxford once gave a lecture in which he asserted that there are many languages in which a double negative makes a positive but none in which a double positive makes a negative — to which the Columbia philosopher Sidney Morgenbesser, sitting in the audience, sarcastically replied, "Yeah, yeah.")

Still, there are plenty of cases where the real world does mirror the rules of negative numbers. One nerve cell's firing can be inhibited by the firing of a second nerve cell. If that second nerve cell is then inhibited by a third, the first cell can fire again. The indirect action of the third cell on the first is tantamount to excitation; a chain of two negatives makes a positive. Similar effects occur in gene regulation: a protein can turn a gene on by blocking another molecule that was repressing that stretch of DNA.

Perhaps the most familiar parallel occurs in the social and political realms as summed up by the adage "The enemy of my enemy is my friend." This truism, and related ones about the friend of my enemy, the enemy of my friend, and so on, can be depicted in relationship triangles.

The corners signify people, companies, or countries, and the sides connecting them signify their relationships, which can be positive (friendly, shown here as solid lines) or negative (hostile, shown as dashed lines).

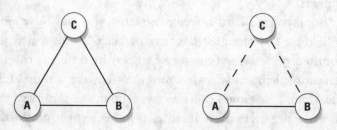

Social scientists refer to triangles like the one on the left, with all sides positive, as balanced—there's no reason for anyone to change how he feels, since it's reasonable to like your friends' friends. Similarly, the triangle on the right, with two negatives and a positive, is considered balanced because it causes no dissonance; even though it allows for hostility, nothing cements a friendship like hating the same person.

Of course, triangles can also be unbalanced. When three mutual enemies size up the situation, two of them—often the two with the least animosity toward each other—may be tempted to join forces and gang up on the third.

Even more unbalanced is a triangle with a single negative relationship. For instance, suppose Carol is friendly with both Alice and Bob, but Bob and Alice despise each other. Perhaps they were once a couple but suffered a nasty breakup, and each is now badmouthing the other to ever-loyal Carol. This causes psychological stress all around. To restore balance, either Alice and Bob have to reconcile or Carol has to choose a side.

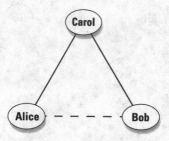

In all these cases, the logic of balance matches the logic of multiplication. In a balanced triangle, the sign of the product of any two sides, positive or negative, always agrees with the sign of the third. In unbalanced triangles, this pattern is broken.

Leaving aside the verisimilitude of the model, there are interesting questions here of a purely mathematical flavor. For example, in a close-knit network where everyone knows everyone, what's the most stable state? One possibility is a nirvana of goodwill, where all relationships are positive and all triangles within the network are balanced. But surprisingly, there are other states that are equally stable. These are states of intractable conflict, with the network split into two hostile factions (of arbitrary sizes and compositions). All members of one faction are friendly with one another but antagonistic toward everybody in the other faction. (Sound familiar?) Perhaps even more surprisingly, these polarized states are the *only* states as stable as nirvana. In particular, no three-party split can have all its triangles balanced.

Scholars have used these ideas to analyze the run-up to World War I. The diagram that follows shows the shifting alliances among Great Britain, France, Russia, Italy, Germany, and Austria-Hungary between 1872 and 1907.

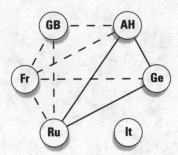

Three Emperors' League
1872–81

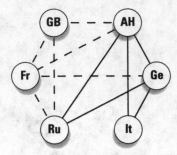

Triple Alliance 1882

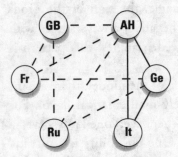

German-Russian Lapse 1890

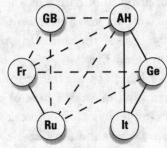

French-Russian Alliance
1891–94

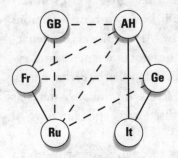

Entente Cordiale 1904

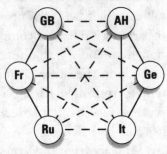

British-Russian Alliance 1907

The first five configurations were all unbalanced, in the sense that they each contained at least one unbalanced triangle. The resultant dissonance tended to push these nations to realign themselves, triggering reverberations elsewhere in the network. In the final stage, Europe had split into two implacably opposed blocs—technically balanced, but on the brink of war.

The point is not that this theory is powerfully predictive. It isn't. It's too simple to account for all the subtleties of geopolitical dynamics. The point is that some part of what we observe is due to nothing more than the primitive logic of "the enemy of my enemy," and *this* part is captured perfectly by the multiplication of negative numbers. By sorting the meaningful from the generic, the arithmetic of negative numbers can help us see where the real puzzles lie.

Commuting

4

EVERY DECADE OR SO a new approach to teaching math comes along and creates fresh opportunities for parents to feel inadequate. Back in the 1960s, my parents were flabbergasted by their inability to help me with my second-grade homework. They'd never heard of base 3 or Venn diagrams.

Now the tables have turned. "Dad, can you show me how to do these multiplication problems?" *Sure*, I thought, until the headshaking began. "No, Dad, that's not how we're supposed to do it. That's the old-school method. Don't you know the lattice method? No? Well, what about partial products?"

These humbling sessions have prompted me to revisit multiplication from scratch. And it's actually quite subtle, once you start to think about it.

Take the terminology. Does "seven times three" mean "seven added to itself three times"? Or "three added to itself seven times"?

In some cultures the language is less ambiguous. A friend of mine from Belize used to recite his times tables like this: "Seven ones are seven, seven twos are fourteen, seven threes are twenty-one," and so on. This phrasing makes it clear that the first number is the multiplier; the second number is the thing being multiplied. It's the same convention as in Lionel Richie's immortal lyrics "She's once, twice, three times a lady." ("She's a lady times three" would never have been a hit.)

Maybe all this semantic fuss strikes you as silly, since the order in which numbers are multiplied doesn't matter anyway: $7 \times 3 = 3 \times 7$. Fair enough, but that begs the question I'd like to explore in some depth here: Is this commutative law of multiplication, $a \times b = b \times a$, really so obvious? I remember being surprised by it as a child; maybe you were too.

To recapture the magic, imagine not knowing what 7×3 equals. So you try counting by sevens: 7, 14, 21. Now turn it around and count by threes instead: 3, 6, 9, . . . Do you feel the suspense building? So far, none of the numbers match those in the sevens list, but keep going . . . 12, 15, 18, and then, bingo, 21!

My point is that if you regard multiplication as being synonymous with repeated counting by a certain number (or, in other words, with repeated addition), the commutative law isn't transparent.

But it becomes more intuitive if you conceive of multiplication *visually*. Think of 7×3 as the number of dots in a rectangular array with seven rows and three columns.

If you turn the array on its side, it transforms into three rows and seven columns—and since rotating the picture doesn't change the number of dots, it must be true that $7 \times 3 = 3 \times 7$.

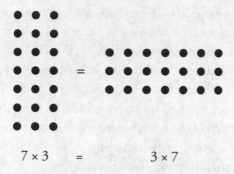

$$7 \times 3 \quad = \quad 3 \times 7$$

Yet strangely enough, in many real-world situations, especially where money is concerned, people seem to forget the commutative law, or don't realize it applies. Let me give you two examples.

Suppose you're shopping for a new pair of jeans. They're on sale for 20 percent off the sticker price of $50, which sounds like a bargain, but keep in mind that you also have to pay the 8 percent sales tax. After the clerk finishes complimenting you on the flattering fit, she starts ringing up the purchase but then pauses and whispers, in a conspiratorial tone, "Hey, let me save you some money. I'll apply the tax first, and then take twenty percent off the total, so you'll get more money back. Okay?"

But something about that sounds fishy to you. "No thanks," you say. "Could you please take the twenty percent off first, then apply the tax to the sale price? That way, I'll pay less tax."

Which way is a better deal for you? (Assume both are legal.)

When confronted with a question like this, many people approach it *additively*. They work out the tax and the discount under both scenarios, and then do whatever additions or subtractions are necessary to find the final price. Doing things the clerk's way, you reason, would cost you $4 in tax (8 percent of the sticker price of $50). That would bring your total to $54.

Then applying the 20 percent discount to $54 gives you $10.80 back, so you'd end up paying $54 minus $10.80, which equals $43.20. Whereas under your scenario, the 20 percent discount would be applied first, saving you $10 off the $50 sticker price. Then the 8 percent tax on that reduced price of $40 would be $3.20, so you'd still end up paying $43.20. Amazing!

But it's merely the commutative law in action. To see why, think *multiplicatively*, not additively. Applying an 8 percent tax followed by a 20 percent discount amounts to multiplying the sticker price by 1.08 and then multiplying that result by 0.80. Switching the order of tax and discount reverses the multiplication, but since $1.08 \times 0.80 = 0.80 \times 1.08$, the final price comes out the same.

Considerations like these also arise in larger financial decisions. Is a Roth 401(k) better or worse than a traditional retirement plan? More generally, if you have a pile of money to invest and you have to pay taxes on it at some point, is it better to take the tax bite at the beginning of the investment period, or at the end?

Once again, the commutative law shows it's a wash, all other things being equal (which, sadly, they often aren't). If, for both scenarios, your money grows by the same factor and gets taxed at the same rate, it doesn't matter whether you pay the taxes up front or at the end.

Please don't mistake this mathematical remark for financial advice. Anyone facing these decisions in real life needs to be aware of many complications that muddy the waters: Do you expect to be in a higher or lower tax bracket when you retire? Will you max out your contribution limits? Do you think the government will change its policies about the tax-exempt status of withdrawals by the time you're ready to take the money out? Leaving all this aside (and don't get me wrong, it's all important; I'm just trying to focus here on a simpler mathematical issue), my basic point is that the commutative law is relevant to the analysis of such decisions.

You can find heated debates about this on personal finance sites on the Internet. Even after the relevance of the commutative law has been pointed out, some bloggers don't accept it. It's that counterintuitive.

Maybe we're wired to doubt the commutative law because in daily life, it usually matters what you do first. You can't have your cake and eat it too. And when taking off your shoes and socks, you've got to get the sequencing right.

The physicist Murray Gell-Mann came to a similar realization one day when he was worrying about his future. As an undergraduate at Yale, he desperately wanted to stay in the Ivy League for graduate school. Unfortunately Princeton rejected his application. Harvard said yes but seemed to be dragging its feet about providing the financial support he needed. His best option, though he found it depressing, was MIT. In Gell-Mann's eyes, MIT was a grubby technological institute, beneath his rarefied taste. Nevertheless, he accepted the offer. Years later he would explain that he had contemplated suicide at the time but decided against it once he realized that attending MIT and killing himself didn't commute. He could always go to MIT and commit suicide later if he had to, but not the other way around.

Gell-Mann had probably been sensitized to the importance of non-commutativity. As a quantum physicist he would have been acutely aware that at the deepest level, nature disobeys the commutative law. And it's a good thing, too. For the failure of commutativity is what makes the world the way it is. It's why matter is solid, and why atoms don't implode.

Specifically, early in the development of quantum mechanics, Werner Heisenberg and Paul Dirac had discovered that nature follows a curious kind of logic in which $p \times q \neq q \times p$, where p and q represent the momentum and position of a quantum particle. Without that breakdown of the commutative law, there would be no Heisenberg uncertainty principle, atoms would collapse, and nothing would exist.

That's why you'd better mind your p's and q's. And tell your kids to do the same.

Division and Its Discontents   5

THERE'S A NARRATIVE line that runs through arithmetic, but many of us missed it in the haze of long division and common denominators. It's the story of the quest for ever more versatile numbers.

The natural numbers 1, 2, 3, and so on are good enough if all we want to do is count, add, and multiply. But once we ask how much remains when everything is taken away, we are forced to create a new kind of number—zero—and since debts can be owed, we need negative numbers too. This enlarged universe of numbers, called integers, is every bit as self-contained as the natural numbers but much more powerful because it embraces subtraction as well.

A new crisis comes when we try to work out the mathematics of sharing. Dividing a whole number evenly is not always possible . . . unless we expand the universe once more, now by inventing fractions. These are ratios of integers—hence their technical name, rational numbers. Sadly, this is the place where many students hit the mathematical wall.

There are many confusing things about division and its consequences, but perhaps the most maddening is that there are so many different ways to describe a part of a whole.

If you cut a chocolate layer cake right down the middle into two equal pieces, you could certainly say that each piece is half the cake. Or you might express the same idea with the fraction 1/2, meaning "1 of 2 equal pieces." (When you write it this way, the slash between the 1 and the 2 is a visual reminder that something is being sliced.) A third way is to say that each piece is 50 percent of the whole, meaning literally "50 parts out of 100." As if that weren't enough, you could also invoke decimal notation and describe each piece as 0.5 of the entire cake.

This profusion of choices may be partly to blame for the bewilderment many of us feel when confronted with fractions, percentages, and decimals. A vivid example appears in the movie *My Left Foot*, the true story of the Irish writer, painter, and poet Christy Brown. Born into a large working-class family, he suffered from cerebral palsy that made it almost impossible for him to speak or control any of his limbs except his left foot. As a boy he was often dismissed as mentally disabled, especially by his father, who resented him and treated him cruelly.

A pivotal scene in the movie takes place around the kitchen table. One of Christy's older sisters is quietly doing her math homework, seated next to her father, while Christy, as usual, is shunted off in the corner of the room, twisted in his chair. His sister breaks the silence: "What's twenty-five percent of a quarter?" she asks. Father mulls it over. "Twenty-five percent of a quarter? Uhhh . . . That's a stupid question, eh? I mean, twenty-five percent *is* a quarter. You can't have a quarter of a quarter." Sister responds, "You can. Can't you, Christy?" Father: "Ha! What would *he* know?"

Writhing, Christy struggles to pick up a piece of chalk with his left foot. Positioning it over a slate on the floor, he manages to scrawl a 1, then a slash, then something unrecognizable. It's the number 16, but the 6 comes out backward. Frustrated, he erases the 6 with his heel and tries again, but this time the chalk moves too far, crossing through the 6, rendering it indecipherable. "That's only a nervous squiggle," sneers his father, turning away. Christy closes his eyes and slumps back, exhausted.

Aside from the dramatic power of the scene, what's striking is the father's conceptual rigidity. What makes him insist you can't have a quarter of a quarter? Maybe he thinks you can take a quarter only out of a whole or from something made of four equal parts. But what he fails to realize is that *everything* is made of four equal parts. In the case of an object that's already a quarter, its four equal parts look like this:

Since 16 of these thin slices make the original whole, each slice is 1/16 of the whole—the answer Christy was trying to scratch out.

A version of the same kind of mental rigidity, updated for the digital age, made the rounds on the Internet a few years ago when a frustrated customer named George Vaccaro recorded and posted his phone conversation with two service representatives at Verizon Wireless. Vaccaro's complaint was that he'd been quoted a data usage rate of .002 cents per kilobyte, but his bill showed he'd been charged .002 *dollars* per kilobyte, a hundredfold higher rate. The ensuing conversation climbed to the top fifty in YouTube's comedy section.

Here's a highlight that occurs about halfway through the recording, during an exchange between Vaccaro and Andrea, the Verizon floor manager:

> v: Do you recognize that there's a difference between one dollar and one cent?
> a: Definitely.
> v: Do you recognize there's a difference between half a dollar and half a cent?
> a: Definitely.
> v: Then, do you therefore recognize there's a difference between .002 dollars and .002 cents?
> a: No.
> v: No?
> a: I mean there's . . . there's no .002 dollars.

A few moments later Andrea says, "Obviously a dollar is 'one, decimal, zero, zero,' right? So what would a 'point zero zero two dollars' look like? . . . I've never heard of .002 dollars . . . It's just not a full cent."

The challenge of converting between dollars and cents is only part of the problem for Andrea. The real barrier is her inability to envision a portion of either.

From firsthand experience I can tell you what it's like to be mystified by decimals. In eighth grade, Ms. Stanton began teaching us how to convert a fraction into a decimal. Using long division, we found that some fractions give decimals that terminate in all zeros. For example, $\frac{1}{4}$ = .2500 . . . , which can be rewritten as .25, since all those zeros amount to nothing. Other fractions give decimals that eventually repeat, like

$$\frac{5}{6} = .8333 \ldots$$

My favorite was $\frac{1}{7}$, whose decimal counterpart repeats every six digits:

$$\frac{1}{7} = .142857142857 \ldots$$

The bafflement began when Ms. Stanton pointed out that if you triple both sides of the simple equation

$$\frac{1}{3} = .3333 \ldots ,$$

you're forced to conclude that 1 must equal .9999 . . .

At the time I protested that they couldn't be equal. No matter how many 9s she wrote, I could write just as many 0s in 1.0000 . . . and then if we subtracted her number from mine, there would be a teeny bit left over, something like .0000 . . . 01.

Like Christy's father and the Verizon service reps, I couldn't accept something that had just been proven to me. I saw it but refused to believe it. (This might remind you of some people you know.)

But it gets worse—or better, if you like to feel your neurons sizzle. Back in Ms. Stanton's class, what stopped us from looking at decimals that neither terminate nor repeat periodically? It's easy to cook up such stomach-churners. Here's an example:

0.12122122212222 . . .

By design, the blocks of 2 get progressively longer as we move to the right. There's no way to express this decimal as a fraction. Fractions always yield decimals that terminate or eventually repeat periodically—that can be proven—and since this decimal does neither, it can't be equal to the ratio of any whole numbers. It's irrational.

Given how contrived this decimal is, you might suppose irrationality is rare. On the contrary, it is typical. In a certain sense that can be made precise, almost all decimals are irrational. And their digits look statistically random.

Once you accept these astonishing facts, everything turns topsy-turvy. Whole numbers and fractions, so beloved and familiar, now appear scarce and exotic. And that innocuous number line pinned to the molding of your grade-school classroom? No one ever told you, but it's chaos up there.

I'D WALKED PAST Ezra Cornell's statue hundreds of times without even glancing at his greenish likeness. But then one day I stopped for a closer look.

Ezra appears outdoorsy and ruggedly dignified in his long coat, vest, and boots, his right hand resting on a walking stick and holding a rumpled, wide-brimmed hat. The monument comes across as unpretentious and disarmingly direct — much like the man himself, by all accounts.

Which is why it seems so discordant that Ezra's dates are inscribed on the pedestal in pompous Roman numerals:

EZRA CORNELL

MDCCCVII–MDCCCLXXIV

Why not write simply 1807–1874? Roman numerals may look impressive, but they're hard to read and cumbersome to use. Ezra would have had little patience for that.

Finding a good way to represent numbers has always been a challenge. Since the dawn of civilization, people have tried various systems for writing numbers and reckoning with them, whether for trading, measuring land, or keeping track of the herd.

What nearly all these systems have in common is that our biology is deeply embedded in them. Through the vagaries of evolution, we happen to have five fingers on each of two hands. That peculiar anatomical fact is reflected in the primitive system of tallying; for example, the number 17 is written as

Here, each of the vertical strokes in each group must have originally meant a finger. Maybe the diagonal slash was a thumb, folded across the other four fingers to make a fist?

Roman numerals are only slightly more sophisticated than tallies. You can spot the vestige of tallies in the way Romans wrote 2 and 3, as II and III. Likewise, the diagonal slash is echoed in the shape of the Roman symbol for 5, V. But 4 is an ambiguous case. Sometimes it's written as IIII, tally style (you'll often see this on fancy clocks), though more commonly it's written as IV. The positioning of a smaller number (I) to the left of a larger number (V) indicates that you're supposed to subtract I, rather than add it, as you would if it were stationed on the right. Thus IV means 4, whereas VI means 6.

The Babylonians were not nearly as attached to their fingers. Their numeral system was based on 60—a clear sign of their impeccable taste, for 60 is an exceptionally pleasant number. Its beauty is intrinsic and has nothing to do with human appendages. Sixty is the smallest number that can be divided evenly by 1, 2, 3, 4, 5, and 6. And that's just for starters (there's also 10, 12, 15, 20, and 30). Because of its promiscuous divisibility, 60 is much more congenial than 10 for any sort of calculation or measurement that involves cutting things into equal parts. When we divide an hour into 60 minutes, or a minute into 60 seconds, or a full circle into 360 degrees, we're channeling the sages of ancient Babylon.

But the greatest legacy of the Babylonians is an idea that's so commonplace today that few of us appreciate how subtle and ingenious it is.

To illustrate it, let's consider our own Hindu-Arabic system, which incorporates the same idea in its modern form. Instead of 60, this system is based on ten symbols: 1, 2, 3, 4, 5, 6,

7, 8, 9, and, most brilliant, 0. These are called digits, naturally, from the Latin word for a finger or a toe.

The great innovation here is that even though this system is based on the number 10, there is no single symbol reserved for 10. Ten is marked by a *position*—the tens place—instead of a symbol. The same is true for 100, or 1,000, or any other power of 10. Their distinguished status is signified not by a symbol but by a parking spot, a reserved piece of real estate. Location, location, location.

Contrast the elegance of this place-value system with the much cruder approach used in Roman numerals. You want 10? We've got 10. It's X. We've also got 100 (C) and 1,000 (M), and we'll even throw in special symbols for the 5 family: V, L, and D, for 5, 50, and 500.

The Roman approach was to elevate a few favored numbers, give them their own symbols, and express all the other, second-class numbers as combinations of those.

Unfortunately, Roman numerals creaked and groaned when faced with anything larger than a few thousand. In a workaround solution that would nowadays be called a kludge, the scholars who were still using Roman numerals in the Middle Ages resorted to piling bars on top of the existing symbols to indicate multiplication by a thousand. For instance, $\overline{X}$ meant ten thousand, and $\overline{M}$ meant a thousand thousands or, in other words, a million. Multiplying by a billion (a thousand million) was rarely necessary, but if you ever had to, you could always put a second bar on top of the $\overline{M}$. As you can see, the fun never stopped.

But in the Hindu-Arabic system, it's a snap to write any number, no matter how big. All numbers can be expressed with the same ten digits, merely by slotting them into the right places. Furthermore, the notation is inherently concise. Every number less than a million, for example, can be expressed in six symbols or fewer. Try doing that with words, tallies, or Roman numerals.

Best of all, with a place-value system, ordinary people can learn to do arithmetic. You just have to master a few facts — the multiplication table and its counterpart for addition. Once you get those down, that's all you'll ever need. Any calculation involving any pair of numbers, no matter how big, can be performed by applying the same sets of facts, over and over again, recursively.

If it all sounds pretty mechanical, that's precisely the point. With place-value systems, you can program a machine to do arithmetic. From the early days of mechanical calculators to the supercomputers of today, the automation of arithmetic was made possible by the beautiful idea of place value.

But the unsung hero in this story is 0. Without 0, the whole approach would collapse. It's the placeholder that allows us to tell 1, 10, and 100 apart.

All place-value systems are based on some number called, appropriately enough, the base. Our system is base 10, or decimal (from the Latin root *decem*, meaning "ten"). After the ones place, the subsequent consecutive places represent tens, hundreds, thousands, and so on, each of which is a power of 10:

$$10 = 10^1$$
$$100 = 10 \times 10 = 10^2$$
$$1,000 = 10 \times 10 \times 10 = 10^3.$$

Given what I said earlier about the biological, as opposed to the logical, origin of our preference for base 10, it's natural to ask: Would some other base be more efficient, or easier to manipulate?

A strong case can be made for base 2, the famous and now ubiquitous binary system used in computers and all things digital, from cell phones to cameras. Of all the possible bases, it requires the fewest symbols — just two of them, 0 and 1. As such, it meshes perfectly with the logic of electronic switches or anything else that can toggle between two states — on or off, open or closed.

Binary takes some getting used to. Instead of powers of 10, it uses powers of 2. It still has a ones place like the decimal system, but the subsequent places now stand for twos, fours, and eights, because

$$2 = 2^1$$
$$4 = 2 \times 2 = 2^2$$
$$8 = 2 \times 2 \times 2 = 2^3.$$

Of course, we wouldn't write the symbol 2, because it doesn't exist in binary, just as there's no single numeral for 10 in decimal. In binary, 2 is written as 10, meaning one 2 and zero 1s. Similarly, 4 would be written as 100 (one 4, zero 2s, and zero 1s), and 8 would be 1000.

The implications reach far beyond math. Our world has been changed by the power of 2. In the past few decades we've come to realize that *all* information — not just numbers, but also language, images, and sound — can be encoded in streams of zeros and ones.

Which brings us back to Ezra Cornell.

Tucked at the rear of his monument, and almost completely obscured, is a telegraph machine — a modest reminder of his role in the creation of Western Union and the tying together of the North American continent.

As a carpenter turned entrepreneur, Cornell worked for Samuel Morse, whose name lives on in the code of dots and dashes through which the English language was reduced to the clicks of a telegraph key. Those two little symbols were technological forerunners of today's zeros and ones.

Morse entrusted Cornell to build the nation's first telegraph line, a link from Baltimore to the U.S. Capitol, in Washington, D.C. From the very start it seems that he had an inkling of what his dots and dashes would bring. When the line was officially opened, on May 24, 1844, Morse sent the first message down the wire: "What hath God wrought."

Part Two **RELATIONSHIPS**

Now it's time to move on from grade-school arithmetic to high-school math. Over the next ten chapters we'll be revisiting algebra, geometry, and trig. Don't worry if you've forgotten them all—there won't be any tests this time around. Instead of worrying about the details of these subjects, we have the luxury of concentrating on their most beautiful, important, and far-reaching ideas.

Algebra, for example, may have struck you as a dizzying mix of symbols, definitions, and procedures, but in the end they all boil down to just two activities—solving for x and working with formulas.

Solving for x is detective work. You're searching for an unknown number, x. You've been handed a few clues about it, either in the form of an equation like $2x + 3 = 7$ or, less conveniently, in a convoluted verbal description of it (as in those scary word problems). In either case, the goal is to identify x from the information given.

Working with formulas, by contrast, is a blend of art and science. Instead of dwelling on a particular x, you're manipulating and massaging relationships that continue to hold even as the numbers in them change. These changing numbers are called variables, and they are what truly distinguishes algebra from arithmetic. The formulas in question might express elegant patterns about numbers for their own sake. This is where algebra meets art. Or they might express relationships between numbers in the real world, as they do in the laws of nature for falling objects or planetary orbits or genetic frequencies in a population. This is where algebra meets science.

This division of algebra into two grand activities is not standard (in fact, I just made it up), but it seems to work pretty well. In the next chapter I'll have more to say about solving for x, so for now let's focus on formulas, starting with some easy examples to clarify the ideas.

A few years ago, my daughter Jo realized something about her big sister, Leah. "Dad, there's always a number between my age and Leah's. Right now I'm six and Leah's eight, and seven is in the middle. And even when we're old, like when I'm twenty and she's twenty-two, there will still be a number in the middle!"

Jo's observation qualifies as algebra (though no one but a proud father would see it that way) because she was noticing a relationship between two ever-changing variables: her age, x, and Leah's age, y. No matter how old both of them are, Leah will always be two years older: $y = x + 2$.

Algebra is the language in which such patterns are most naturally phrased. It takes some practice to become fluent in algebra, because it's loaded with what the French call *faux amis*, "false friends": a pair of words, each from a different language (in this case, English and algebra), that sound related and seem to mean the same thing but that actually mean something horribly different from each other when translated.

For example, suppose the length of a hallway is y when

measured in yards, and f when measured in feet. Write an equation that relates y to f.

My friend Grant Wiggins, an education consultant, has been posing this problem to students and faculty for years. He says that in his experience, students get it wrong more than half the time, even if they have recently taken and passed an algebra course.

If you think the answer is $y = 3f$, welcome to the club.

It seems like such a straightforward translation of the sentence "One yard equals three feet." But as soon as you try a few numbers, you'll see that this formula gets everything backward. Say the hallway is 10 yards long; everyone knows that's 30 feet. Yet when you plug in $y = 10$ and $f = 30$, the formula doesn't work!

The correct formula is $f = 3y$. Here 3 really means "3 feet per yard." When you multiply it by y in yards, the units of yards cancel out and you're left with units of feet, as you should be.

Checking that the units cancel properly helps avoid this kind of blunder. For example, it could have saved the Verizon customer service reps (discussed in chapter 5) from confusing dollars and cents.

Another kind of formula is known as an identity. When you factored or multiplied polynomials in algebra class, you were working with identities. You can use them now to impress your friends with numerical parlor tricks. Here's one that impressed the physicist Richard Feynman, no slouch himself at mental math:

> When I was at Los Alamos I found out that Hans Bethe was absolutely topnotch at calculating. For ex-

ample, one time we were putting some numbers into a formula, and got to 48 squared. I reach for the Marchant calculator, and he says, "That's 2,300." I begin to push the buttons, and he says, "If you want it exactly, it's 2,304."

The machine says 2,304. "Gee! That's pretty remarkable!" I say.

"Don't you know how to square numbers near 50?" he says. "You square 50—that's 2,500—and subtract 100 times the difference of your number from 50 (in this case it's 2), so you have 2,300. If you want the correction, square the difference and add it on. That makes 2,304."

Bethe's trick is based on the identity

$$(50 + x)^2 = 2,500 + 100x + x^2.$$

He had memorized that equation and was applying it for the case where x is -2, corresponding to the number $48 = 50 - 2$.

For an intuitive proof of this formula, imagine a square patch of carpet that measures $50 + x$ on each side.

	50	x
50	2500	$50x$
x	$50x$	x^2

Then its area is $(50 + x)$ squared, which is what we're looking for. But the diagram above shows that this area is made of a 50-by-50 square (this contributes the 2,500 to the formula), two rectangles of dimensions 50 by x (each contributes an area of 50x, for a combined total of 100x), and finally the little x-by-x square gives an area of x squared, the final term in Bethe's formula.

Relationships like these are not just for theoretical physicists. An identity similar to Bethe's is relevant to anyone who has money invested in the stock market. Suppose your portfolio drops catastrophically by 50 percent one year and then gains 50 percent the next. Even after that dramatic recovery, you'd still be down 25 percent. To see why, observe that a 50 percent loss multiplies your money by 0.50, and a 50 percent gain multiplies it by 1.50. When those happen back to back, your money multiplies by 0.50 times 1.50, which equals 0.75 — in other words, a 25 percent loss.

In fact, you *never* get back to even when you lose and gain by the same percentage in consecutive years. With algebra we can understand why. It follows from the identity

$$(1 - x)(1 + x) = 1 - x^2.$$

In the down year the portfolio shrinks by a factor $1 - x$ (where $x = 0.50$ in the example above), and then grows by a factor $1 + x$ the following year. So the net change is a factor of

$$(1 - x)(1 + x)$$

and according to the formula above, this equals

$$1 - x^2.$$

The point is that this expression is *always* less than 1 for any x other than 0. So you never completely recoup your losses.

Needless to say, not every relationship between variables is as straightforward as those above. Yet the allure of algebra is seductive, and in gullible hands it spawns such silliness as a formula for the socially acceptable age difference in a romance. According to some sites on the Internet, if your age is x, polite society will disapprove if you date someone younger than $x/2 + 7$.

In other words, it would be creepy for anyone over eighty-two to eye my forty-eight-year-old wife, even if she were available. But eighty-one? No problem.

Ick. Ick. Ick . . .

Finding Your Roots

8

FOR MORE THAN 2,500 years, mathematicians have been obsessed with solving for x. The story of their struggle to find the roots — the solutions — of increasingly complicated equations is one of the great epics in the history of human thought.

One of the earliest such problems perplexed the citizens of Delos around 430 B.C. Desperate to stave off a plague, they consulted the oracle of Delphi, who advised them to double the volume of their cube-shaped altar to Apollo. Unfortunately, it turns out that doubling a cube's volume required them to construct the cube root of 2, a task that is now known to be impossible, given their restriction to use nothing but a straightedge and compass, the only tools allowed in Greek geometry.

Later studies of similar problems revealed another irritant, a nagging little thing that wouldn't go away: even when solutions were possible, they often involved square roots of negative numbers. Such solutions were long derided as sophistic or fictitious because they seemed nonsensical on their face.

Until the 1700s or so, mathematicians believed that square roots of negative numbers simply couldn't exist.

They couldn't be positive numbers, after all, since a positive times a positive is always positive, and we're looking for numbers whose square is negative. Nor could negative numbers work, since a negative times a negative is, again, *positive*. There seemed to be no hope of finding numbers that when multiplied by themselves would give negative answers.

We've seen crises like this before. They occur whenever an existing operation is pushed too far, into a domain where it no longer seems sensible. Just as subtracting bigger numbers from smaller ones gave rise to negative numbers (chapter 3) and division spawned fractions and decimals (chapter 5), the freewheeling use of square roots eventually forced the universe of numbers to expand . . . again.

Historically, this step was the most painful of all. The square root of −1 still goes by the demeaning name of i, for "imaginary."

This new kind of number (or if you'd rather be agnostic, call it a symbol, not a number) is defined by the property that

$$i^2 = -1.$$

It's true that i can't be found anywhere on the number line. In that respect it's much stranger than zero, negative numbers, fractions, or even irrational numbers, all of which—weird as they are—still have their places in line.

But with enough imagination, our minds can make room for i as well. It lives off the number line, at right angles to it, on its own imaginary axis. And when you fuse that imaginary axis to the ordinary "real" number line, you create a 2-D space—a plane—where a new species of numbers lives.

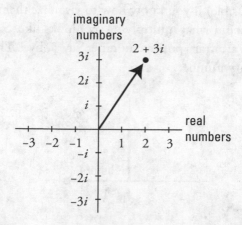

These are the complex numbers. Here "complex" doesn't mean "complicated"; it means that two types of numbers, real and imaginary, have bonded together to form a complex, a hybrid number like $2 + 3i$.

Complex numbers are magnificent, the pinnacle of number systems. They enjoy all the same properties as real numbers—you can add and subtract them, multiply and divide them—but they are *better* than real numbers because they always have roots. You can take the square root or cube root or any root of a complex number, and the result will still be a complex number.

Better yet, a grand statement called the fundamental theorem of algebra says that the roots of any polynomial are always complex numbers. In that sense they're the end of the quest, the holy grail. The universe of numbers need never expand again. Complex numbers are the culmination of the journey that began with 1.

You can appreciate the utility of complex numbers (or find

it more plausible) if you know how to visualize them. The key is to understand what multiplying by *i* looks like. Suppose we multiply an arbitrary positive number, say 3, by *i*. The result is the imaginary number 3*i*.

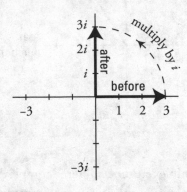

So multiplying by *i* produces a rotation counterclockwise by a quarter turn. It takes an arrow of length 3 pointing east and changes it into a new arrow of the same length but now pointing north.

Electrical engineers love complex numbers for exactly this reason. Having such a compact way to represent 90-degree rotations is very useful when working with alternating currents and voltages, or with electric and magnetic fields, because these often involve oscillations or waves that are a quarter cycle (i.e., 90 degrees) out of phase.

In fact, complex numbers are indispensable to all engineers. In aerospace engineering they eased the first calculations of the lift on an airplane wing. Civil and mechanical engineers use them routinely to analyze the vibrations of footbridges, skyscrapers, and cars driving on bumpy roads.

The 90-degree rotation property also sheds light on what $i^2 = -1$ really means. If we multiply a positive number by i^2, the corresponding arrow rotates 180 degrees, flipping from east to west, because the two 90-degree rotations (one for each factor of i) combine to make a 180-degree rotation.

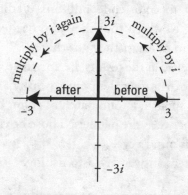

But multiplying by -1 produces the very same 180-degree flip. That's the sense in which $i^2 = -1$.

Computers have breathed new life into complex numbers and the age-old problem of root finding. When they're not being used for Web surfing or e-mail, the machines on our desks can reveal things the ancients never dreamed of.

In 1976, my Cornell colleague John Hubbard began looking at the dynamics of Newton's method, a powerful algorithm for finding roots of equations in the complex plane. The method takes a starting point (an approximation to the root) and does a certain computation that improves it. By doing this repeatedly, always using the previous point to generate a better one, the method bootstraps its way forward and rapidly homes in on a root.

Hubbard was interested in problems with *multiple* roots. In that case, which root would the method find? He proved that if there were just two roots, the closer one would always win.

But if there were three or more roots, he was baffled. His earlier proof no longer applied.

So Hubbard did an experiment. A *numerical* experiment.

He programmed a computer to run Newton's method. Then he told it to color-code millions of different starting points according to which root they approached and to shade them according to how fast they got there.

Before he peeked at the results, he anticipated that the roots would most quickly attract the points nearby and thus should appear as bright spots in a solid patch of color. But what about the boundaries between the patches? Those he couldn't picture, at least not in his mind's eye.

The computer's answer was astonishing.

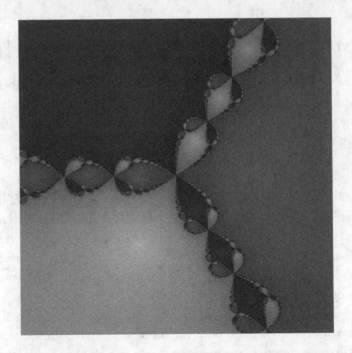

The borderlands looked like psychedelic hallucinations. The colors intermingled there in an almost impossibly promiscuous manner, touching each other at infinitely many points and always in a three-way. In other words, wherever two colors met, the third would always insert itself and join them.

Magnifying the boundaries revealed patterns within patterns.

The structure was a fractal—an intricate shape whose inner structure repeated at finer and finer scales.

Furthermore, chaos reigned near the boundary. Two points might start very close together, bounce side by side for a while, and then veer off to different roots. The winning root was as unpredictable as the winning number in a game of roulette. Little things—tiny, imperceptible changes in the initial conditions—could make all the difference.

Hubbard's work was an early foray into what's now called complex dynamics, a vibrant blend of chaos theory, complex analysis, and fractal geometry. In a way it brought geometry back to its roots. In 600 B.C. a manual written in Sanskrit for temple builders in India gave detailed geometric instructions for computing square roots, needed in the design of ritual altars. More than 2,500 years later, in 1976, mathematicians were still searching for roots, but now the instructions were written in binary code.

Some imaginary friends you never outgrow.

My Tub Runneth Over

9

UNCLE IRV WAS my dad's brother as well as his partner in a shoe store they owned in our town. He handled the business end of things and mostly stayed in the office upstairs, because he was good with numbers and not so good with the customers.

When I was about ten or eleven, Uncle Irv gave me my first word problem. It sticks with me to this day, probably because I got it wrong and felt embarrassed.

It had to do with filling a bathtub. If the cold-water faucet can fill the tub in a half-hour, and the hot-water faucet can fill it in an hour, how long will it take to fill the tub when they're running together?

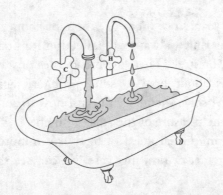

I'm pretty sure I guessed forty-five minutes, as many people would. Uncle Irv shook his head and grinned. Then, in his high-pitched nasal voice, he proceeded to school me.

"Steven," he said, "figure out how much water pours into the tub in a minute." The cold water fills the tub in thirty minutes, so in one minute it fills $\frac{1}{30}$ of the tub. But the hot water runs slower—it takes sixty minutes, which means it fills only $\frac{1}{60}$ of the tub per minute. So when they're both running, they fill

$$\frac{1}{30} + \frac{1}{60}$$

of the tub in a minute.

To add those fractions, observe that 60 is their lowest common denominator. Then, rewriting $\frac{1}{30}$ as $\frac{2}{60}$, we get

$$\frac{1}{30} + \frac{1}{60} = \frac{2}{60} + \frac{1}{60}$$
$$= \frac{3}{60}$$
$$= \frac{1}{20},$$

which means that the two faucets working together fill $\frac{1}{20}$ of the tub per minute. So they fill the whole tub in twenty minutes.

Over the years since then, I've thought about this bathtub problem many times, always with affection for both Uncle Irv and the question itself. There are broader lessons to be learned here—lessons about how to solve problems approximately when you can't solve them exactly, and how to solve them intuitively, for the pleasure of the Aha! moment.

Consider my initial guess of forty-five minutes. By looking at an extreme, or limiting, case, we can see that that answer can't possibly be right. In fact, it's absurd. To understand why, suppose the hot water wasn't turned on. Then the cold water — on its own — would fill the tub in thirty minutes. So whatever the answer to Uncle Irv's question is, it has to be less than this. After all, running the hot water along with the cold can only help.

Admittedly, this conclusion is not as informative as the exact answer of twenty minutes we found by Uncle Irv's method, but it has the advantage of not requiring any calculation.

A different way to simplify the problem is to pretend the two faucets run at the same rate. Say each can fill the tub in thirty minutes (meaning that the hot water runs just as fast as the cold). Then the answer would be obvious. Because of the symmetry of the new situation, the two perfectly matched faucets would together fill the tub in fifteen minutes, since each does half the work.

This instantly tells us that Uncle Irv's scenario must take longer than fifteen minutes. Why? Because fast plus fast beats slow plus fast. Our make-believe symmetrical problem has two fast faucets, whereas Uncle Irv's has one slow, one fast. And since fifteen minutes is the answer when they're both fast, Uncle Irv's tub can only take longer.

The upshot is that by considering two hypothetical cases — one with the hot water off, and another with it matched to the cold water — we learned that the answer lies somewhere between fifteen and thirty minutes. In much harder problems where it may be impossible to find an exact answer — not just in math but in other domains as well — this sort of partial information can be very valuable.

Even if we're lucky enough to come up with an exact answer, that's still no cause for complacency. There may be easier or clearer ways to find the solution. This is one place where math allows for creativity.

For example, instead of Uncle Irv's textbook method, with its fractions and common denominators, here's a more playful route to the same result. It dawned on me some years later, when I tried to pinpoint why the problem is so confusing in the first place and realized it's because of the faucets' different speeds. That makes it a headache to keep track of what each faucet contributes, especially if you picture the hot and cold water sloshing together and mixing in the tub.

So let's keep the two types of water apart, at least in our minds. Instead of a single bathtub, imagine two assembly lines of them, two separate conveyor belts shuttling bathtubs past a hot-water faucet on one side and a cold-water faucet on the other.

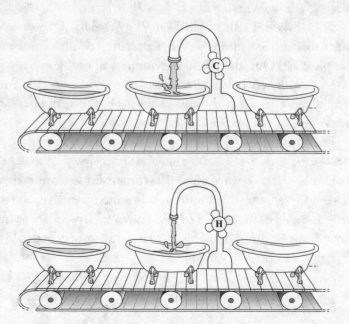

Each faucet stands in place and fills its own tubs — no mixing allowed. And as soon as a tub fills up, it moves on down the line, making way for the next one.

Now everything becomes easy. In one hour, the hot-water faucet fills one tub, while the cold-water faucet fills two (since each takes a half-hour). That amounts to three tubs per hour, or one tub every twenty minutes. Eureka!

So why do so many people, including my childhood self, blunder into guessing forty-five minutes? Why is it so tempting to split the difference of thirty and sixty minutes? I'm not sure, but it seems to be a case of faulty pattern recognition. Maybe the bathtub problem is being conflated with others where splitting the difference would make sense. My wife explained it to me by analogy. Imagine you're helping a little old lady cross the street. Without your help, it would take her sixty seconds, while you'd zip across in thirty seconds. How long, then, would it take the two of you, walking arm in arm? A compromise around forty-five seconds seems plausible because when granny is clinging to your elbow, she slows you down and you speed her up.

The difference here is that you and granny affect each other's speeds, but the faucets don't. They're independent. Apparently our subconscious minds don't spot this distinction, at least not when they're leaping to the wrong conclusion.

The silver lining is that even wrong answers can be educational . . . as long as you realize they're wrong. They expose misguided analogies and other woolly thinking, and bring the crux of the problem into sharper relief.

Other classic word problems are expressly designed to trick their victims by misdirection, like a magician's sleight of hand. The phrasing of the question sets a trap. If you answer by instinct, you'll probably fall for it.

Try this one. Suppose three men can paint three fences in three hours. How long would it take one man to paint one fence?

It's tempting to blurt out "one hour." The words themselves nudge you that way. The drumbeat in the first sentence—three men, three fences, three hours—catches your attention by establishing a rhythm, so when the next sentence repeats the pattern with one man, one fence, ____ hours, it's hard to resist filling in the blank with "one." The parallel construction suggests an answer that's linguistically right but mathematically wrong.

The correct answer is three hours.

If you *visualize* the problem—mentally picture three men painting three fences and all finishing after three hours, just as the problem states—the right answer becomes clear. For all three fences to be done after three hours, each man must have spent three hours on his.

The undistracted reasoning that this problem requires is one of the most valuable things about word problems. They force us to pause and think, often in unfamiliar ways. They give us practice in being mindful.

Perhaps even more important, word problems give us practice in thinking not just about numbers, but about *relationships* between numbers—how the flow rates of the faucets affect the time required to fill the tub, for example. And that is the essential next step in anyone's math education. Understandably, a lot of us have trouble with it; relationships are much more abstract than numbers. But they're also much more powerful. They express the inner logic of the world around us. Cause and effect, supply and demand, input and output, dose and response—all involve pairs of numbers and the relationships between them. Word problems initiate us into this way of thinking.

However, Keith Devlin raises an interesting criticism in his essay "The problem with word problems." His point is that these problems typically assume you understand the rules of the game and agree to play by them, even though they're often artificial, sometimes absurdly so. For example, in our problem about three men painting three fences in three hours, it was implicit that (1) all three men paint at the same rate and (2) they all paint steadily, never slowing down or speeding up. Both assumptions are unrealistic. You're supposed to know not to worry about that and go along with the gag, because otherwise the problem would be too complicated and you wouldn't have enough information to solve it. You'd need to know exactly how much each painter slows down as he gets tired in the third hour, how often each one stops for a snack, and so on.

Those of us who teach math should try to turn this bug into a feature. We should be up front about the fact that word problems force us to make simplifying assumptions. That's a valuable skill—it's called mathematical modeling. Scientists do it all the time when they apply math to the real world. But they, unlike the authors of most word problems, are usually careful to state their assumptions explicitly.

So thanks, Uncle Irv, for that first lesson. Humiliating? Yes. Unforgettable? Yes, that too . . . but in a good way.

THE QUADRATIC FORMULA is the Rodney Dangerfield of algebra. Even though it's one of the all-time greats, it don't get no respect.

Professionals certainly aren't enamored of it. When mathematicians and physicists are asked to list the top ten most beautiful or important equations of all time, the quadratic formula never makes the cut. Oh sure, everybody swoons over $1 + 1 = 2$, and $E = mc^2$, and the pert little Pythagorean theorem, strutting like it's all that just because $a^2 + b^2 = c^2$. But the quadratic formula? Not a chance.

Admittedly, it's unsightly. Some students prefer to sound it out, treating it as a ritual incantation: "x equals negative b, plus or minus the square root of b squared minus four a c, all over two a." Others made of sterner stuff look the formula straight in the face, confronting a hodgepodge of letters and symbols more formidable than anything they've encountered up to that point:

$$x = \frac{-b \pm \sqrt{b^2 - 4ac}}{2a}.$$

It's only when you understand what the quadratic formula is trying to do that you can begin to appreciate its inner beauty. In this chapter I hope to give you a feeling for the cleverness packed into that porcupine of symbols, along with a better sense of what the formula means and where it arises.

There are many situations in which we'd like to figure out the value of some unknown number. What dose of radiation therapy should be given to shrink a thyroid tumor? How much money would you have to pay each month to cover a thirty-year mortgage of $200,000 at a fixed annual interest rate of 5 percent? How fast does a rocket have to go to escape the Earth's gravity?

Algebra is the place where we cut our teeth on the simplest problems of this type. The subject was developed by Islamic mathematicians around A.D. 800, building on earlier work by Egyptian, Babylonian, Greek, and Indian scholars. One practical impetus at that time was the challenge of calculating inheritances according to Islamic law.

For example, suppose a widower dies and leaves his entire estate of 10 dirhams to his daughter and two sons. Islamic law requires that both the sons must receive equal shares. Moreover, each son must receive twice as much as the daughter. How many dirhams will each heir receive?

Let's use the letter x to denote the daughter's inheritance. Even though we don't know what x is yet, we can reason about it as if it were an ordinary number. Specifically, we know that each son gets twice as much as the daughter does, so they each receive $2x$. Thus, taken together, the amount that the three heirs inherit is $x + 2x + 2x$, for a total of $5x$, and this must equal the total value of the estate, 10 dirhams. Hence $5x = 10$ dirhams. Finally, by dividing both sides of the equation by 5, we see that $x = 2$ dirhams is the daughter's share. And since each of the sons inherits $2x$, they both get 4 dirhams.

Notice that two types of numbers appeared in this problem: known numbers, like 2, 5, and 10, and unknown numbers, like x. Once we managed to derive a relationship between the unknown and the known (as encapsulated in the equation $5x = 10$), we were able to chip away at the equation, dividing both sides by 5 to isolate the unknown x. It was a bit like a sculptor working the marble, trying to release the statue from the stone.

A slightly different tactic would have been needed if we had encountered a known number being *subtracted* from an unknown, as in an equation like $x - 2 = 5$. To free x in this case, we would pare away the 2 by adding it to both sides of the equation. This yields an unencumbered x on the left and $5 + 2 = 7$ on the right. Thus $x = 7$, which you may have already realized by common sense.

Although this tactic is now familiar to all students of algebra, they may not realize the entire subject is named after it. In the early part of the ninth century, Muhammad ibn Musa al-Khwarizmi, a mathematician working in Baghdad, wrote a seminal textbook in which he highlighted the usefulness of restoring a quantity being subtracted (like 2, above) by adding it to the other side of an equation. He called this process *al-jabr* (Arabic for "restoring"), which later morphed into "algebra." Then, long after his death, he hit the etymological jackpot again. His own name, al-Khwarizmi, lives on today in the word "algorithm."

In his textbook, before wading into the intricacies of calculating inheritances, al-Khwarizmi considered a more complicated class of equations that embody relationships among *three* kinds of numbers, not the mere two considered above. Along with known numbers and an unknown (x), these equations also included the square of the unknown (x^2). They are now called quadratic equations, from the Latin *quadratus*, for "square." Ancient scholars in Babylonia, Egypt, Greece, China, and India had already tackled such brainteasers, which often arose in architectural or geometrical problems involving areas or proportions, and had shown how to solve some of them.

An example discussed by al-Khwarizmi is

$$x^2 + 10x = 39.$$

In his day, however, such problems were posed in words, not symbols. He asked: "What must be the square which, when increased by ten of its own roots, amounts to thirty-nine?" (Here, the term "root" refers to the unknown x.)

This problem is much tougher than the two we considered above. How can we isolate x now? The tricks used earlier are insufficient, because the x^2 and $10x$ terms tend to step on each other's toes. Even if you manage to isolate x in one of them, the other remains troublesome. For instance, if we divide both sides of the equation by 10, the $10x$ simplifies to x, which is what we want, but then the x^2 becomes $x^2/10$, which brings us no closer to finding x itself. The basic obstacle, in a nutshell, is that we have to do two things at once, and they seem almost incompatible.

The solution that al-Khwarizmi presents is worth delving into in some detail, first because it's so slick, and second because it's so powerful—it allows us to solve *all* quadratic equations in a single stroke. By that I mean that if the known numbers 10 and 39 above were changed to any other numbers, the method would still work.

The idea is to interpret each of the terms in the equation geometrically. Think of the first term, x^2, as the area of a square with dimensions x by x.

$$\begin{array}{c} x \\ x\ \boxed{\ x^2\ } \end{array}$$

Similarly, regard the second term, $10x$, as the area of a rectangle of dimensions 10 by x or, more ingenious, as the area of two equal rectangles, each measuring 5 by x. (Splitting the rectangle into two pieces sets the stage for the key maneuver that follows, known as completing the square.)

Attach the two new rectangles onto the square to produce a notched shape of area $x^2 + 10x$:

Viewed in this light, al-Khwarizmi's puzzle amounts to asking: If the notched shape occupies 39 square units of area, how large would x have to be?

$$x^2 + 10x = 39$$

The picture itself suggests an almost irresistible next step. Look at that missing corner. If only it were filled in, the notched shape would turn into a perfect square. So let's take the hint and complete the square.

$$(x + 5)^2 = 64$$

Supplying the missing 5×5 square adds 25 square units to the existing area of $x^2 + 10x$, for a total of $x^2 + 10x + 25$. Equivalently, that combined area can be expressed more neatly as $(x + 5)^2$, since the completed square is $x + 5$ units long on each side.

The upshot is that x^2 and $10x$ are now moving gracefully as a couple, rather than stepping on each other's toes, by being paired within the single expression $(x + 5)^2$. That's what will soon enable us to solve for x.

Meanwhile, because we added 25 units of area to the left side of the equation $x^2 + 10x = 39$, we must also add 25 to the right side, to keep the equation balanced. Since $39 + 25 = 64$, our equation then becomes

$$(x + 5)^2 = 64.$$

But that's a cinch to solve. Taking square roots of both sides gives $x + 5 = 8$, so $x = 3$.

Lo and behold, 3 really does solve the equation $x^2 + 10x = 39$. If we square 3 (giving 9) and then add 10 times 3 (giving 30), the sum is 39, as desired.

There's only one snag. If al-Khwarizmi were taking algebra today, he wouldn't receive full credit for this answer. He fails to mention that a negative number, $x = -13$, also works. Squaring it gives 169; adding it ten times gives −130; and they too add up to 39. But this negative solution was ignored in ancient times, since a square with a side of negative length is geometrically meaningless. Today, algebra is less beholden to geometry and we regard the positive and negative solutions as equally valid.

In the centuries after al-Khwarizmi, scholars came to realize that *all* quadratic equations could be solved in the same way, by completing the square—as long as one was willing to allow the negative numbers (and their bewildering square roots) that often came up in the answers. This line of argument revealed that the solutions to any quadratic equation

$$ax^2 + bx + c = 0$$

(where a, b, and c are known but arbitrary numbers, and x is unknown) could be expressed by the quadratic formula,

$$x = \frac{-b \pm \sqrt{b^2 - 4ac}}{2a}.$$

What's so remarkable about this formula is how brutally explicit and comprehensive it is. There's the answer, right there, no matter what a, b, and c happen to be. Considering that there are infinitely many possible choices for each of them, that's a lot for a single formula to manage.

In our own time, the quadratic formula has become an irreplaceable tool for practical applications. Engineers and scientists use it to analyze the tuning of a radio, the swaying of a footbridge or a skyscraper, the arc of a baseball or a cannonball, the ups and downs of animal populations, and countless other real-world phenomena.

For a formula born of the mathematics of inheritance, that's quite a legacy.

IF YOU WERE an avid television watcher in the 1980s, you may remember a clever show called *Moonlighting*. Known for its snappy dialogue and the romantic chemistry between its costars, it featured Cybill Shepherd and Bruce Willis as Maddie Hayes and David Addison, a couple of wisecracking private detectives.

While investigating one particularly tough case, David asks a coroner's assistant for his best guess about possible suspects. "Beats me," says the assistant. "But you know what I don't understand?" David replies, "Logarithms?" Then, reacting to Maddie's look: "What? You understood those?"

That pretty well sums up how many people feel about logarithms. Their peculiar name is just part of their image problem. Most folks never use them again after high school, at least not consciously, and are oblivious to the logarithms hiding behind the scenes of their daily lives.

The same is true of many of the other functions discussed in algebra II and precalculus. Power functions, exponential functions—what was the point of all that? My goal in this chapter is to help you appreciate the function of all those functions, even if you never have occasion to press their buttons on your calculator.

A mathematician needs functions for the same reason that a builder needs hammers and drills. Tools transform things. So do functions. In fact, mathematicians often refer to them as transformations because of this. But instead of wood and steel, the materials that functions pound away on are numbers and shapes and, sometimes, even other functions.

To show you what I mean, let's plot the graph of the equation $y = 4 - x^2$. You may remember how this sort of activity goes: You draw a picture of the xy plane with the x-axis running horizontally and the y-axis vertically. Then for each x you compute the corresponding y and plot them together as a single point in the xy plane. For example, when x is 1, the equation says $y = 4 - 1^2$, which is $4 - 1$, or 3. So $(x,y) = (1,3)$ is a point on the graph. After you calculate and plot a few more points, the following picture emerges.

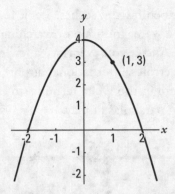

The bowed shape of the curve is due to the action of mathematical pliers. In the equation for y, the function that transforms x into x^2 behaves a lot like the common tool for bending and pulling things. When it's applied to every point on a piece of the x-axis (which you could visualize as a straight piece of wire), the pliers bend and elongate that piece into the downward-curving arch shown above.

And what role does the 4 play in the equation $y = 4 - x^2$? It acts like a nail for hanging a picture on a wall. It lifts up the bent wire arch by 4 units. Since it raises all points by the same amount, it's known as a constant function.

This example illustrates the dual nature of functions. On the one hand, they're tools: the x^2 bends the piece of the x-axis, and the 4 lifts it. On the other hand, they're building blocks: the 4 and the $-x^2$ can be regarded as component parts of a more complicated function, $4 - x^2$, just as wires, batteries, and transistors are component parts of a radio.

Once you start to look at things this way, you'll notice functions everywhere. The arching curve above—technically known as a parabola—is the signature of the squaring function x^2 operating behind the scenes. Look for it when you're taking a sip from a water fountain or watching a basketball arc toward the hoop. And if you ever have a few minutes to spare on a layover in Detroit's international airport, be sure to stop by the water feature in the Delta terminal to enjoy the world's most breathtaking parabolas at play.

Parabolas and constants are associated with a wider class of functions—power functions of the form x^n, in which a variable x is raised to a fixed power n. For a parabola, $n = 2$; for a constant, $n = 0$.

Changing the value of n yields other handy tools. For example, raising x to the first power ($n = 1$) gives a function that works like a ramp, a steady incline of growth or decay. It's called a linear function because its xy graph is a line. If you leave a bucket out in a steady rain, the water collecting at the bottom rises linearly in time.

Another useful tool is the inverse square function, $1/x^2$, corresponding to the case $n = -2$. (The power becomes -2 because the function is an *inverse* square; the x^2 appears in the denominator.) This function is good for describing how waves and forces attenuate as they spread out in three dimensions—for instance, how a sound softens as it moves away from its source.

Power functions like these are the building blocks that scientists and engineers use to describe growth and decay in their mildest forms.

But when you need mathematical dynamite, it's time to unpack the exponential functions. They describe all sorts of explosive growth, from nuclear chain reactions to the proliferation of bacteria in a petri dish. The most familiar example is the function 10^x, in which 10 is raised to the power x. Make sure not to confuse this with the earlier power functions. Here the exponent (the power x) is a variable, and the base (the number 10) is a constant—whereas in a power function like x^2, it's the other way around. This switch makes a huge difference: as x gets larger and larger, an exponential function of x eventually grows faster than *any* power function, no matter how large the power. Exponential growth is almost unimaginably rapid.

That's why it's so hard to fold a piece of paper in half more than seven or eight times. Each folding approximately doubles the thickness of the wad, causing it to grow exponentially. Meanwhile, the wad's length shrinks in half every time, and thus *decreases* exponentially fast. For a standard sheet of notebook paper, after seven folds the wad becomes thicker than it is long, so it can't be folded again. It doesn't matter how strong the person doing the folding is. For a sheet to be considered legitimately folded n times, the resulting wad is required to have 2^n layers in a straight line, and this can't happen if the wad is thicker than it is long.

The challenge was thought to be impossible until Britney Gallivan, then a junior in high school, solved it in 2002. She began by deriving a formula

$$L = \frac{\pi T}{6}(2^n + 4)(2^n - 1)$$

that predicted the maximum number of times, n, that paper of a given thickness T and length L could be folded in one direction. Notice the forbidding presence of the exponential function 2^n in two places — once to account for the doubling of the wad's thickness at each fold, and another time to account for the halving of its length.

Using her formula, Britney concluded that she would need to use a special roll of toilet paper nearly three-quarters of a mile long. She bought the paper, and in January 2002, she went to a shopping mall in her hometown of Pomona, California, and unrolled the paper. Seven hours later, and with the help of her parents, she smashed the world record by folding the paper in half twelve times!

In theory, exponential growth is also supposed to grace your bank account. If your money grows at an annual interest rate of r, after one year it will be worth $(1 + r)$ times your original deposit; after two years, $(1 + r)^2$; and after x years, $(1 + r)^x$ times your initial deposit. Thus the miracle of compounding that we so often hear about is caused by exponential growth in action.

Which brings us back to logarithms. We need them because it's useful to have tools that can undo the actions of other tools. Just as every office worker needs both a stapler and a staple remover, every mathematician needs exponential functions *and* logarithms. They're inverses. This means that if you type a number x into your calculator and then punch the 10^x button followed by the log x button, you'll get back to the number you started with. For example, if $x = 2$, then 10^x would be 10^2, which equals 100. Taking the log of that then brings the result back to 2; the log button undoes the action of the 10^x button. Hence log(100) equals 2. Likewise, log(1,000) = 3 and log(10,000) = 4, because 1,000 = 10^3 and 10,000 = 10^4.

Notice something magical here: as the numbers inside the logarithms grew *multiplicatively*, increasing tenfold each time from 100 to 1,000 to 10,000, their logarithms grew *additively*, increasing from 2 to 3 to 4. Our brains perform a similar trick when we listen to music. The frequencies of the notes in a scale—do, re, mi, fa, sol, la, ti, do—sound to us like they're rising in equal *steps*. But objectively their vibrational frequencies are rising by equal *multiples*. We perceive pitch logarithmically.

In every place where they arise, from the Richter scale for earthquake magnitudes to pH measures of acidity, logarithms make wonderful compressors. They're ideal for taking quantities that vary over a wide range and squeezing them together so they become more manageable. For instance, 100 and 100 million differ a millionfold, a gulf that most of us find incomprehensible. But their logarithms differ only fourfold (they are 2 and 8, because 100 = 10^2 and 100 million = 10^8). In conversation, we all use a crude version of logarithmic shorthand when we refer to any salary between $100,000 and $999,999 as being six figures. That "six" is roughly the logarithm of these salaries, which in fact span the range from five to six.

As impressive as all these functions may be, a mathematician's toolbox can only do so much—which is why I still haven't assembled my Ikea bookcases.

Part Three **SHAPES**

I BET I CAN guess your favorite math subject in high school.

It was geometry.

So many people I've met over the years have expressed affection for that subject. Is it because geometry draws on the right side of the brain, and that appeals to visual thinkers who might otherwise cringe at its cold logic? Maybe. But some people tell me they loved geometry precisely because it *was* so logical. The step-by-step reasoning, with each new theorem resting firmly on those already established—that's the source of satisfaction for many.

But my best hunch (and, full disclosure, I personally love geometry) is that people enjoy it because it *marries* logic and intuition. It feels good to use both halves of the brain.

To illustrate the pleasures of geometry, let's revisit the Pythagorean theorem, which you probably remember as $a^2 + b^2 = c^2$. Part of the goal here is to see why it's true and appreciate why it matters. Beyond that, by proving the theorem in two different ways, we'll come to see how one proof can be more elegant than another, even though both are correct.

The Pythagorean theorem is concerned with right triangles—meaning those with a right (90-degree) angle at one of

the corners. Right triangles are important because they're what
you get if you cut a rectangle in half along its diagonal:

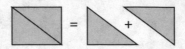

And since rectangles come up often in all sorts of settings, so
do right triangles.

They arise, for instance, in surveying. If you're measuring
a rectangular field, you might want to know how far it is from
one corner to the diagonally opposite corner. (By the way, this
is where geometry started, historically—in problems of land
measurement, or measuring the earth: *geo* = "earth," and *metry*
= "measurement.")

The Pythagorean theorem tells you how long the diagonal
is compared to the sides of the rectangle. If one side has length
a and the other has length *b*, the theorem says the diagonal has
length *c*, where

$$a^2 + b^2 = c^2.$$

For some reason, the diagonal is traditionally called the
hypotenuse, though I've never met anyone who knows why.
(Any Latin or Greek scholars?) It must have something to do
with the diagonal subtending a right angle, but as jargon goes,
"subtending" is about as opaque as "hypotenuse."

Anyway, here's how the theorem works. To keep the numbers simple, let's say $a = 3$ yards and $b = 4$ yards. Then to figure out the unknown length c, we don our black hoods and intone that c^2 is the sum of 3^2 plus 4^2, which is 9 plus 16. (Keep in mind that all of these quantities are now measured in square yards, since we squared the yards as well as the numbers themselves.) Finally, since $9 + 16 = 25$, we get $c^2 = 25$ square yards, and then taking square roots of both sides yields $c = 5$ yards as the length of the hypotenuse.

This way of looking at the Pythagorean theorem makes it seem like a statement about lengths. But traditionally it was viewed as a statement about *areas*. That becomes clearer when you hear how they used to say it:

> The square on the hypotenuse is the sum of the squares on the other two sides.

Notice the word "on." We're not speaking of the square *of* the hypotenuse — that's a newfangled algebraic concept about multiplying a number (the length of the hypotenuse) by itself. No, we're referring here to a square literally sitting *on* the hypotenuse, like this:

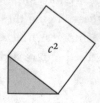

Let's call this the large square, to distinguish it from the small and medium squares we can build on the other two sides:

Then the theorem says that the large square has the same area as the small and medium squares combined.

For thousands of years, this marvelous fact has been expressed in a diagram, an iconic mnemonic of dancing squares:

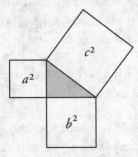

Viewing the theorem in terms of areas makes it a lot more fun to think about. For example, you can test it — and then eat it — by building the squares out of many little crackers. Or you can treat the theorem like a child's puzzle, with pieces of different shapes and sizes. By rearranging these puzzle pieces, we can prove the theorem very simply, as follows.

Let's go back to the tilted square sitting on the hypotenuse.

At an instinctive level, you should feel a bit disquieted by this image. The square looks potentially unstable, like it might topple or slide down the ramp. And there's also an unpleasant arbitrariness about which of the four sides of the square gets to touch the triangle.

Guided by these intuitive feelings, let's add three more copies of the triangle around the square to make a more solid and symmetrical picture:

Now recall what we're trying to prove: that the tilted white square in the picture above (which is just our earlier large square—it's still sitting right there on the hypotenuse) has the same area as the small and medium squares put together. But where are those other squares? Well, we have to shift some triangles around to find them.

Think of the picture above as depicting a puzzle, with four triangular pieces wedged into the corners of a rigid puzzle frame.

In this interpretation, the tilted square is the empty space in the middle of the puzzle. The rest of the area inside the frame is occupied by the puzzle pieces.

Now let's try moving the pieces around in various ways. Of course, nothing we do can ever change the total amount of empty space inside the frame — it's always whatever area lies outside the pieces.

The brainstorm, then, is to rearrange the pieces like this:

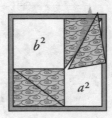

All of a sudden the empty space has changed into the two shapes we're looking for — the small square and the medium square. And since the total area of empty space always stays the same, we've just proven the Pythagorean theorem!

This proof does far more than convince; it *illuminates*. That's what makes it elegant.

For comparison, here's another proof. It's equally famous, and it's perhaps the simplest proof that avoids using areas.

As before, consider a right triangle with sides of length a and b and hypotenuse of length c, as shown below on the left.

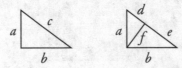

Now, by divine inspiration or a stroke of genius, something tells us to draw a line segment perpendicular to the hypotenuse and down to the opposite corner, as shown in the triangle on the right.

This clever little construction creates two smaller triangles inside the original one. It's easy to prove that all these triangles are similar — which means they have identical shapes but different sizes. That in turn implies that the lengths of their corresponding parts have the same proportions, which translates into the following set of equations:

$$\frac{a}{f} = \frac{b}{e} = \frac{c}{b}$$

$$\frac{a}{d} = \frac{b}{f} = \frac{c}{a}$$

We also know that

$$c = d + e$$

because our construction merely split the original hypotenuse of length c into two smaller sides of lengths d and e.

At this point you might be feeling a bit lost, or at least unsure of what to do next. There's a morass of five equations above, and we're trying to whittle them down to deduce that

$$a^2 + b^2 = c^2.$$

Try it for a few minutes. You'll discover that two of the equations are irrelevant. That's ugly; an elegant proof should involve nothing superfluous. With hindsight, of course, you wouldn't have listed those equations to begin with. But that would just be putting lipstick on a p . . . (the missing word here is "proof").

Nevertheless, by manipulating the right three equations, you can get the theorem to pop out. See the notes on page 272 for the missing steps.

Would you agree with me that, on aesthetic grounds, this proof is inferior to the first one? For one thing, it drags near the end. And who invited all that algebra to the party? This is supposed to be a geometry event.

But a more serious defect is the proof's murkiness. By the time you're done slogging through it, you might believe the theorem (grudgingly), but you still might not *see* why it's true.

Leaving proofs aside, why does the Pythagorean theorem even matter? Because it reveals a fundamental truth about the nature of space. It implies that space is flat, as opposed to curved. For the surface of a globe or a bagel, for example, the theorem needs to be modified. Einstein confronted this challenge in his general theory of relativity (where gravity is no longer viewed as a force, but rather as a manifestation of the curvature of space), and so did Bernhard Riemann and others before him when laying the foundations of non-Euclidean geometry.

It's a long road from Pythagoras to Einstein. But at least it's a straight line . . . for most of the way.

Something from Nothing

EVERY MATH COURSE contains at least one notoriously difficult topic. In arithmetic, it's long division. In algebra, it's word problems. And in geometry, it's proofs.

Most students who take geometry have never seen a proof before. The experience can come as a shock, so perhaps a warning label would be in order, something like this: *Proofs can cause dizziness or excessive drowsiness. Side effects of prolonged exposure may include night sweats, panic attacks, and, in rare cases, euphoria. Ask your doctor if proofs are right for you.*

Disorienting as proofs can be, learning to do them has long been thought essential to a liberal education — more essential than the subject matter itself, some would say. According to this view, geometry is good for the mind; it trains you to think clearly and logically. It's not the study of triangles, circles, and parallel lines per se that matters. What's important is the axiomatic method, the process of building a rigorous argument, step by step, until a desired conclusion has been established.

Euclid laid down this deductive approach in the *Elements* (now the most reprinted textbook of all time) about 2,300 years ago. Ever since, Euclidean geometry has been a model for logical reasoning in all walks of life, from science and law to philosophy and politics. For example, Isaac Newton channeled Euclid in the structure of his masterwork *The Mathematical Principles of Natural Philosophy*. Using geometrical proofs, he deduced Galileo's and Kepler's laws about projectiles and planets from his own deeper laws of motion and gravity. Spinoza's *Ethics* follows the same pattern. Its full title is *Ethica Ordine Geometrico Demonstrata* (*Ethics Demonstrated in Geometrical Order*). You can even hear echoes of Euclid in the Declaration of Independence. When Thomas Jefferson wrote, "We hold these truths to be self-evident," he was mimicking the style of the *Elements*. Euclid had begun with the definitions, postulates, and self-evident truths of geometry — the axioms — and from them erected an edifice of propositions and demonstrations, one truth linked to the next by unassailable logic. Jefferson organized the Declaration in the same way so that its radical conclusion, that the colonies had the right to govern themselves, would seem as inevitable as a fact of geometry.

If that intellectual legacy seems far-fetched, keep in mind that Jefferson revered Euclid. A few years after he finished his second term as president and stepped out of public life, he wrote to his old friend John Adams on January 12, 1812, about the pleasures of leaving politics behind: "I have given up newspapers in exchange for Tacitus and Thucydides, for Newton and Euclid; and I find myself much the happier."

Still, what's missing in all this worship of Euclidean rationality is an appreciation of geometry's more intuitive aspects. Without inspiration, there'd be no proofs — or theorems to prove in the first place. Like composing music or writing poetry, geometry requires making something from nothing. How does a poet find the right words, or a composer a haunting melody? This is the mystery of the muse, and it's no less mysterious in math than in the other creative arts.

As an illustration, consider the problem of constructing an equilateral triangle—a triangle with all three sides the same length. The rules of the game are that you're given one side of the triangle, the line segment shown here:

Your task is to use that segment, somehow, to construct the other two sides, and to prove that they each have the same length as the original. The only tools at your disposal are a straightedge and a compass. A straightedge allows you to draw a straight line of any length, or to draw a straight line between any two points. A compass allows you to draw a circle of any radius, centered at any point.

Keep in mind, however, that a straightedge is not a ruler. It has no markings on it and can't be used to measure lengths. (Specifically, you can't use it to copy or measure the original segment.) Nor can a compass serve as a protractor; all it can do is draw circles, not measure angles.

Ready? Begin.

This is the moment of paralysis. Where to start?

Logic won't help you here. Skilled problem solvers know that a better approach is to relax and play with the puzzle, hoping to get a feel for it. For instance, maybe it would help to use the straightedge to draw tilted lines through the ends of the segment, like this:

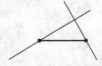

No luck. Although the lines form a triangle, there's no guarantee it's an *equilateral* triangle.

Another stab in the dark would be to draw some circles with the compass. But where? Around one of the endpoints?

Or around a point inside the segment?

That second choice looks unpromising, since there's no reason to pick one interior point over another.

So let's go back to drawing circles around endpoints.

Unfortunately there's still a lot of arbitrariness here. How big should the circles be? Nothing's popping out at us yet.

After a few more minutes of flailing around like this, frustration (and an impending headache) may tempt us to give up. But if we don't, we might get lucky and realize there's only one natural circle to draw. Let's see what happens if we put the sharp point of the compass at one end of the segment and the pencil at the other, and then twirl the compass through a full circle. We'd get this:

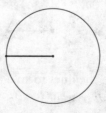

Of course, if we'd used the other endpoint as the center, we'd get this:

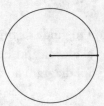

How about drawing both circles at the same time—for no good reason, just noodling?

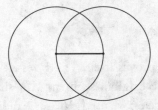

Did it just hit you? A shiver of insight? Look at the diagram again. There's a curvy version of an equilateral triangle staring at us. Its top corner is the point where the circles intersect.

So now let's turn that into a real triangle, with straight sides, by drawing lines from the intersection point to the endpoints of the original segment. The resulting triangle sure looks equilateral.

Having allowed intuition to guide us this far, now and only now is it time for logic to take over and finish the proof. For clarity, let's pan back to the full diagram and label the points of interest *A*, *B*, and *C*.

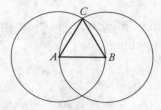

The proof almost proves itself. The sides *AC* and *BC* have the same length as the original segment *AB*, since that's how we constructed the circles; we used *AB* as the radius for both. Since *AC* and *BC* are also radii, they too have this length, so all three lengths are equal, and the triangle is equilateral. QED.

This argument has been around for centuries. In fact, it's Euclid's opening shot — the first proposition in Book I of the *Elements*. But the tendency has always been to present the final diagram with the artful circles already in place, which robs the student of the joy of discovering them. That's a pedagogical mistake. This is a proof that anyone can find. It can be new with each generation, if we teach it right.

The key to this proof, clearly, was the inspired construction of the two circles. A more famous result in geometry can be proven by a similarly deft construction. It's the theorem that the angles of a triangle add up to 180 degrees.

In this case, the best proof is not Euclid's but an earlier one attributed to the Pythagoreans. It goes like this. Consider any triangle, and call its angles *a*, *b*, *c*.

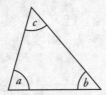

Draw a line parallel to the base, going through the top corner.

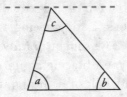

Now we need to digress for a second to recall a property of parallel lines: if another line cuts across two parallel lines like so,

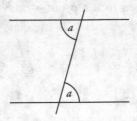

the angles labeled here as *a* (known in the trade as alternate interior angles) are equal.

Let's apply this fact to the construction above in which we drew a line through the top corner of a triangle parallel to its base.

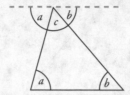

By invoking the equality of alternate interior angles, we see that the angle just to the left of the top corner must be equal to *a*. Likewise the angle at the top right is equal to *b*. So the angles *a*, *b*, and *c* together form a straight angle—an angle of 180 degrees—which is what we sought to prove.

This is one of the most bracing arguments in all of mathematics. It opens with a bolt of lightning, the construction of the parallel line. Once that line has been drawn, the proof practically rises off the table and walks by itself, like Dr. Frankenstein's creation.

And who knows? If we highlight this other side of geometry—its playful, intuitive side, where a spark of imagination can be quickly fanned into a proof—maybe someday all students will remember geometry as the class where they learned to be logical *and* creative.

WHISPERING GALLERIES ARE remarkable acoustic spaces found beneath certain domes, vaults, or curved ceilings. A famous one is located outside the Oyster Bar restaurant in New York City's Grand Central Station. It's a fun place to take a date: the two of you can exchange sweet nothings while you're forty feet apart and separated by a bustling passageway. You'll hear each other clearly, but the passersby won't hear a word you're saying.

To produce this effect, the two of you should stand at diagonally opposite corners of the space, facing the wall. That puts you each near a focus, a special point at which the sound of your voice gets focused as it reflects off the passageway's curved walls and ceiling. Ordinarily, the sound waves you produce travel in all directions and bounce off the walls at disparate times and places, scrambling them so much that they are inaudible when they arrive at the ear of a listener forty feet away (which is why the passersby can't hear what you're saying). But when you whisper at a *focus*, the reflected waves all arrive at the *same* time at the other focus, thus reinforcing one another and allowing your words to be heard.

Ellipses display a similar flair for focusing, though in a much simpler form. If we fashion a reflector in the shape of an ellipse, two particular points inside it (marked as F_1 and F_2 in the figure below) will act as foci in the following sense: all the rays emanating from a light source at one of those points will be reflected to the other.

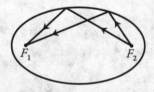

Let me try to convey how amazing that is by restating it in a few ways.

Suppose that Darth and Luke enjoy playing laser tag in an elliptical arena with mirrored walls. Both have agreed not to aim directly at the other—they have to zap each other with bank shots. Darth, not being particularly astute about geometry or optics, suggests they each stand at a focal point. "Fine," says

Luke, "as long as I get to take the first shot." Well, it wouldn't be much of a duel, because Luke can't miss! No matter how foolishly he aims his laser, he'll always tag Darth. Every shot's a winner.

Or if pool is your game, imagine playing billiards on an elliptical table with a pocket at one focus. To set up a trick shot that is guaranteed to scratch every time, place the cue ball at the *other* focus. No matter how you strike the ball and no matter where it caroms off the cushion, it'll always go into the pocket.

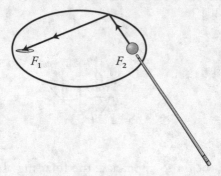

Parabolic curves and surfaces have an impressive focusing power of their own: each can take parallel incoming waves and focus them at a *single* point. This feature of their geometry has been very useful in settings where light waves, sound waves, or other signals need to be amplified. For instance, parabolic microphones can be used to pick up hushed conversations and are therefore of interest in surveillance, espionage, and law enforcement. They also come in handy in nature recording, to capture birdsongs or animal grunts, and in televised sports, to listen in on a coach cursing at an official. Parabolic antennas can amplify radio waves in the same way, which is why satellite

dishes for TV reception and giant radio telescopes for astronomy have their characteristically curved shapes.

This focusing property of parabolas is just as useful when
deployed in reverse. Suppose you want to make a strongly di-
rectional beam of light, like that needed for spotlights or car
headlights. On its own, a bulb—even a powerful one—typically wouldn't be good enough; it wastes too much light by
shining it in all directions. But now place that bulb at the focus
of a parabolic reflector, and voilà! The parabola creates a di-
rectional beam automatically. It takes the bulb's rays and, by
reflecting them off the parabola's silvered inner surface, makes
them all parallel.

Once you begin to appreciate the focusing abilities of parabolas and ellipses, you can't help but wonder if something
deeper is at work here. Are these curves related in some more
fundamental way?

Mathematicians and conspiracy theorists have this much in
common: we're suspicious of coincidences—especially convenient ones. There are no accidents. Things happen for a reason.
While this mindset may be just a touch paranoid when applied
to real life, it's a perfectly sane way to think about math. In the
ideal world of numbers and shapes, strange coincidences usually *are* clues that we're missing something. They suggest the
presence of hidden forces at work.

So let's look more deeply into the possible links between

parabolas and ellipses. At first glance they seem an unlikely couple. Parabolas are arch-shaped and expansive, stretching out at both ends. Ellipses are oval-shaped, like squashed circles, closed and compact.

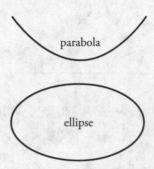

But as soon as you move beyond their appearances and probe their inner anatomy, you start to notice how similar they are. They both belong to a royal family of curves, a genetic tie that becomes obvious — once you know what to look for.

To explain how they're related, we have to recall what, exactly, these curves are.

A parabola is commonly defined as the set of all points equidistant from a given point and a given line not containing that point. That's a mouthful of a definition, but it's actually pretty easy to understand once you translate it into pictures. Call the given point F, for "focus," and call the line L.

$F \bullet$

_____ L

Now, according to the definition, a parabola consists of all the points that lie just as far from F as they do from L. For example, the point P lying straight down from F, halfway to L, would certainly qualify:

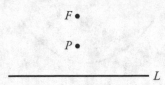

Infinitely many other points P_1, P_2, . . . work too. They lie off to either side like this:

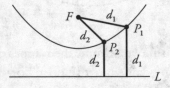

Here, the point P_1 lies at the same distance, d_1, from the line as it does from the focus. The same is true for the point P_2, except now the shared distance is some other number, d_2. Taken together, all the points P with this property form the parabola.

The reason for calling F the focus becomes clear if we think of the parabola as a curved mirror. It turns out (though I won't prove it) that if you shine a beam of light straight at a parabolic mirror, all the reflected rays will intersect simultaneously at the point F, producing an intensely focused spot of light.

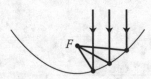

It works something like those old suntanning reflectors that fried a generation of faces, back in the days before anybody worried about skin cancer.

Now let's turn to the corresponding story for ellipses. An ellipse is defined as the set of points the sum of whose distances from two given points is a constant. When rephrased in more down-to-earth language, this description provides a recipe for drawing an ellipse. Get a pencil, a sheet of paper, a corkboard, two pushpins, and a piece of string. Lay the paper on the board. Pin the ends of the string down through the paper, being careful to leave some slack. Then pull the string taut with the pencil to form an angle, as shown below. Begin drawing, keeping the string taut. After the pencil has gone all the way around both pins and returned to its starting point, the resulting curve is an ellipse.

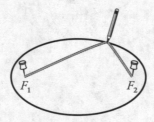

Notice how this recipe implements the definition above, word for word. The pins play the role of the two given points. And the sum of the distances from them to a point on the curve always remains constant, no matter where the pencil is — because those distances always add up to the length of the string.

So where are the ellipse's foci in this construction? At the pins. Again, I won't prove it, but those are the points that allow

Luke and Darth to play their game of can't-miss laser tag and that give rise to a scratch-every-time pool table.

Parabolas and ellipses: Why is it that they, and only they, have such fantastic powers of focusing? What's the secret they share?

They're both cross-sections of the surface of a cone.

A cone? If you feel like that just came out of nowhere, that's precisely the point. The cone's role in all this has been hidden so far.

To see how it's implicated, imagine slicing through a cone with a meat cleaver, somewhat like cutting through a salami, at progressively steeper angles. If the cut is level, the curve of intersection is a circle.

circle

If instead the cone is sliced on a gentle bias, the resulting curve becomes an ellipse.

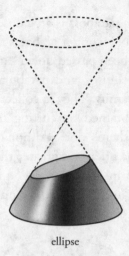

ellipse

As the cuts become steeper, the ellipse gets longer and slimmer in its proportions. At a critical angle, when the bias gets so steep that it matches the slope of the cone itself, the ellipse turns into a parabola.

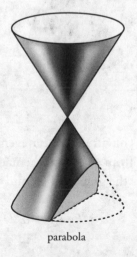

parabola

So that's the secret: a parabola is an ellipse in disguise, in a certain limiting sense. No wonder it shares the ellipse's marvelous focusing ability. It's been passed down through the bloodline.

In fact, circles, ellipses, and parabolas are all members of a larger, tight-knit family. They're collectively known as conic sections — curves obtained by cutting the surface of a cone with a plane. Plus there's one additional sibling: if the cone is sliced very steeply, on a bias greater than its own slope, the resulting incision forms a curve called a hyperbola. Unlike all the others, it comes in two pieces, not one.

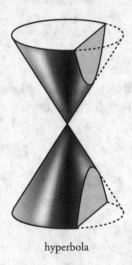

hyperbola

These four types of curves appear even more intimately related when viewed from other mathematical perspectives. In algebra, they arise as the graphs of second-degree equations

$$Ax^2 + Bxy + Cy^2 + Dx + Ey + F = 0,$$

where the constants A, B, C, ... determine whether the graph is a circle, ellipse, parabola, or hyperbola. In calculus, they arise as trajectories of objects tugged by the force of gravity.

So it's no accident that planets move in elliptical orbits with the sun at one focus; or that comets sail through the solar system on elliptic, parabolic, or hyperbolic trajectories; or that a child's ball tossed to a parent follows a parabolic arc. They're all manifestations of the conic conspiracy.

Focus on that the next time you play catch.

Sine Qua Non

My DAD HAD a friend named Dave who retired to Jupiter, Florida. We visited him for a family vacation when I was about twelve or so, and something he showed us made an indelible impression on me.

Dave liked to chart the times of the glorious sunrises and sunsets he could watch from his deck all year long. Every day he marked two dots on his chart, and after many months he noticed something curious. The two curves looked like opposing waves. One usually rose while the other fell; when sunrise was earlier, sunset was later.

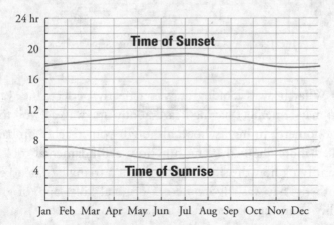

But there were exceptions. For the last three weeks of June and for most of December and early January, sunrise and sunset both came later each day, giving the waves a slightly lopsided appearance.

Still, the message of the curves was unmistakable: the oscillating gap between them showed the days growing longer and shorter with the changing of the seasons. By subtracting the lower curve from the upper one, Dave also figured out how the number of hours of daylight varied throughout the year. To his amazement, *this* curve wasn't lopsided at all. It was beautifully symmetrical.

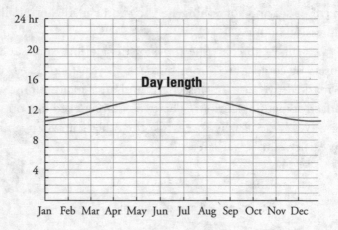

What he was looking at was a nearly perfect sine wave. You may remember having heard about sine waves if you took trigonometry in high school, although your teacher probably talked more about the sine function, a fundamental tool for quantifying how the sides and angles of a triangle relate to one another. That was trigonometry's original killer app, of great utility to ancient astronomers and surveyors.

Yet *tri*gonometry, belying its much too modest name, now goes far beyond the measurement of *tri*angles. By quantifying circles as well, it has paved the way for the analysis of anything that repeats, from ocean waves to brain waves. It's the key to the mathematics of cycles.

To see how trig connects circles, triangles, and waves, imagine a little girl going round and round on a Ferris wheel.

She and her mom, who both happen to be mathematically inclined, have decided this is a perfect opportunity for an experiment. The girl takes a GPS gadget with her on the ride to record her altitude, moment by moment, as the wheel carries her up, then thrillingly over the top and back toward the ground, then up and around again, and so on. The results look like this:

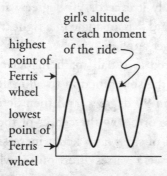

girl's altitude
at each moment
of the ride

highest
point of
Ferris →
wheel

lowest
point of
Ferris →
wheel

This shape is a sine wave. It arises whenever one tracks the horizontal or vertical excursions of something—or someone—moving in a circle.

How is this sine wave related to the sine function discussed in trig class? Well, suppose we examine a snapshot of the girl. At the moment shown, she's at some angle, call it a, relative to the dashed line in the diagram.

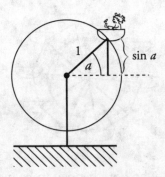

For convenience, let's suppose the hypotenuse of the right triangle shown—which is also the wheel's radius—is 1 unit long. Then sin a (pronounced "sine of a") tells us how high the girl is. More precisely, sin a is defined as the girl's altitude measured from the center of the wheel when she's located at the angle a.

As she goes round and round, her angle a will progressively increase. Eventually it exceeds 90 degrees, at which point we can no longer regard a as an angle in a right triangle. Does that mean trig no longer applies?

No. Undeterred as usual, mathematicians simply enlarge the definition of the sine function to allow for *any* angle, not just those less than 90 degrees, and then define "sin a" as the girl's height above or below the center of the circle. The corresponding graph of sin a, as a keeps increasing (or even goes negative, if the wheel reverses direction), is what we mean by a sine wave. It repeats itself every time *a* changes by 360 degrees, corresponding to a full revolution.

This same sort of conversion of circular motion into sine waves is a pervasive, though often unnoticed, part of our daily experience. It creates the hum of the fluorescent lights overhead in our offices, a reminder that somewhere in the power grid, generators are spinning at sixty cycles per second, converting their rotary motion into alternating current, the electrical sine waves on which modern life depends. When you speak and I hear, both our bodies are using sine waves—yours in the vibrations of your vocal cords to produce the sounds, and mine in the swaying of the hair cells in my ears to receive them. If we open our hearts to these sine waves and tune in to their silent thrumming, they have the power to move us. There's something almost spiritual about them.

When a guitar string is plucked or when children jiggle a jump rope, the shape that appears is a sine wave. The ripples on a pond, the ridges of sand dunes, the stripes of a zebra—all are manifestations of nature's most basic mechanism of pattern formation: the emergence of sinusoidal structure from a background of bland uniformity.

There are deep mathematical reasons for this. Whenever a state of featureless equilibrium loses stability — for whatever reason, and by whatever physical, biological, or chemical process — the pattern that appears first is a sine wave, or a combination of them.

Sine waves are the atoms of structure. They're nature's building blocks. Without them there'd be nothing, giving new meaning to the phrase "sine qua non."

In fact, the words are literally true. Quantum mechanics describes real atoms, and hence all of matter, as packets of sine waves. Even at the cosmological scale, sine waves form the seeds of all that exists. Astronomers have probed the spectrum (the pattern of sine waves) of the cosmic microwave background and found that their measurements match the predictions of inflationary cosmology, the leading theory for the birth and growth of the universe. So it seems that out of a featureless Big

Bang, primordial sine waves—ripples in the density of matter and energy—emerged spontaneously and spawned the stuff of the cosmos.

Stars. Galaxies. And, ultimately, little kids riding Ferris wheels.

IN MIDDLE SCHOOL my friends and I enjoyed chewing on the classic conundrums. What happens when an irresistible force meets an immovable object? Easy — they both explode. Philosophy's trivial when you're thirteen.

But one puzzle bothered us: If you keep moving halfway to the wall, will you ever get there? Something about this one was deeply frustrating, the thought of getting closer and closer and yet never quite making it. (There's probably a metaphor for teenage angst in there somewhere.) Another concern was the thinly veiled presence of infinity. To reach the wall you'd need to take an infinite number of steps, and by the end they'd become infinitesimally small. Whoa.

Questions like this have always caused headaches. Around 500 B.C., Zeno of Elea posed four paradoxes about infinity that puzzled his contemporaries and that may have been partly to blame for infinity's banishment from mathematics for centuries thereafter. In Euclidean geometry, for example, the only constructions allowed were those that involved a finite number of steps. The infinite was considered too ineffable, too unfathomable, and too hard to make logically rigorous.

But Archimedes, the greatest mathematician of antiquity, realized the power of the infinite. He harnessed it to solve problems that were otherwise intractable and in the process came close to inventing calculus—nearly 2,000 years before Newton and Leibniz.

In the coming chapters we'll delve into the great ideas at the heart of calculus. But for now I'd like to begin with the first beautiful hints of them, visible in ancient calculations about circles and pi.

Let's recall what we mean by "pi." It's a ratio of two distances. One of them is the diameter, the distance *across* the circle through its center. The other is the circumference, the distance *around* the circle. "Pi" is defined as their ratio, the circumference divided by the diameter.

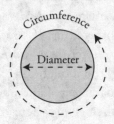

If you're a careful thinker, you might be worried about something already. How do we know that pi is the same number for all circles? Could it be different for big circles and little circles? The answer is no, but the proof isn't trivial. Here's an intuitive argument.

Imagine using a photocopier to reduce an image of a circle by, say, 50 percent. Then *all* distances in the picture—including the circumference and the diameter—would shrink in proportion by 50 percent. So when you divide the new circumference by the new diameter, that 50 percent change would cancel out, leaving the ratio between them unaltered. That ratio is pi.

Of course, this doesn't tell us how big pi is. Simple experiments with strings and dishes are good enough to yield a value near 3, or if you're more meticulous, $3\frac{1}{7}$. But suppose we want to find pi exactly or at least approximate it to any desired accuracy. What then? This was the problem that confounded the ancients.

Before turning to Archimedes's brilliant solution, we should mention one other place where pi appears in connection with circles. The area of a circle (the amount of space inside it) is given by the formula

$$A = \pi r^2.$$

Here A is the area, π is the Greek letter pi, and r is the radius of the circle, defined as half the diameter.

All of us memorized this formula in high school, but where does it come from? It's not usually proven in geometry class. If you went on to take calculus, you probably saw a proof of it there, but is it really necessary to use calculus to obtain something so basic?

Yes, it is.

What makes the problem difficult is that circles are round. If they were made of straight lines, there'd be no issue. Finding the areas of triangles and squares is easy. But working with curved shapes like circles is hard.

The key to thinking mathematically about curved shapes is to pretend they're made up of lots of little straight pieces. That's

not really true, but it works . . . as long as you take it to the
limit and imagine *infinitely* many pieces, each infinitesimally
small. That's the crucial idea behind all of calculus.

Here's one way to use it to find the area of a circle. Begin by
chopping the area into four equal quarters, and rearrange them
like so.

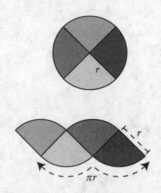

The strange scalloped shape on the bottom has the same area as
the circle, though that might seem pretty uninformative since
we don't know its area either. But at least we know two important facts about it. First, the two arcs along its bottom have a
combined length equal to half the circumference of the original
circle (because the other half of the circumference is accounted
for by the two arcs on top). Since the whole circumference is
pi times the diameter, half of it is pi times *half* the diameter
or, equivalently, pi times the radius r. That's why the diagram
above shows πr as the combined length of the arcs at the bottom of the scalloped shape. Second, the straight sides of the
slices have a length of r, since each of them was originally a
radius of the circle.

Next, repeat the process, but this time with eight slices,
stacked alternately as before.

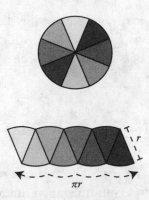

The scalloped shape looks a bit less bizarre now. The arcs on the top and the bottom are still there, but they're not as pronounced. Another improvement is the left and right sides of the scalloped shape don't tilt as much as they used to. Despite these changes, the two facts above continue to hold: the arcs on the bottom still have a net length of πr and each side still has a length of r. And of course the scalloped shape still has the same area as before — the area of the circle we're seeking — since it's just a rearrangement of the circle's eight slices.

As we take more and more slices, something marvelous happens: the scalloped shape approaches a rectangle. The arcs become flatter and the sides become almost vertical.

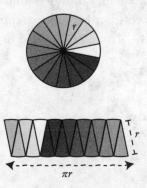

In the limit of *infinitely* many slices, the shape *is* a rectangle. Just as before, the two facts still hold, which means this rectangle has a bottom of width πr and a side of height r.

But now the problem is easy. The area of a rectangle equals its width times its height, so multiplying πr times r yields an area of πr^2 for the rectangle. And since the rearranged shape always has the same area as the circle, that's the answer for the circle too!

What's so charming about this calculation is the way infinity comes to the rescue. At every finite stage, the scalloped shape looks weird and unpromising. But when you take it to the limit—when you finally get to the wall—it becomes simple and beautiful, and everything becomes clear. That's how calculus works at its best.

Archimedes used a similar strategy to approximate pi. He replaced a circle with a polygon with many straight sides and then kept doubling the number of sides to get closer to perfect roundness. But rather than settling for an approximation of uncertain accuracy, he methodically bounded pi by sandwiching the circle between inscribed and circumscribed polygons, as shown below for 6-, 12-, and 24-sided figures.

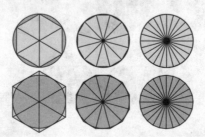

Then he used the Pythagorean theorem to work out the perimeters of these inner and outer polygons, starting with the hexagon and bootstrapping his way up to 12, 24, 48, and, ultimately, 96 sides. The results for the 96-gons enabled him to prove that

$$3\tfrac{10}{71} < \pi < 3\tfrac{1}{7}.$$

In decimal notation (which Archimedes didn't have), this means pi is between 3.1408 and 3.1429.

This approach is known as the method of exhaustion because of the way it traps the unknown number pi between two known numbers that squeeze it from either side. The bounds tighten with each doubling, thus exhausting the wiggle room for pi.

In the limit of infinitely many sides, both the upper and lower bounds would converge to pi. Unfortunately, this limit isn't as simple as the earlier one, where the scalloped shape morphed into a rectangle. So pi remains as elusive as ever. We can discover more and more of its digits—the current record is over 2.7 trillion decimal places—but we'll never know it completely.

Aside from laying the groundwork for calculus, Archimedes taught us the power of approximation and iteration. He bootstrapped a good estimate into a better one, using more and more straight pieces to approximate a curved object with increasing accuracy.

More than two millennia later, this strategy matured into the modern field of numerical analysis. When engineers use computers to design cars that are optimally streamlined, or when biophysicists simulate how a new chemotherapy drug latches onto a cancer cell, they are using numerical analysis.

The mathematicians and computer scientists who pioneered this field have created highly efficient, repetitive algorithms, running billions of times per second, that enable computers to solve problems in every aspect of modern life, from biotech to Wall Street to the Internet. In each case, the strategy is to find a series of approximations that converge to the correct answer as a limit.

And there's no limit to where that'll take us.

Part Four **CHANGE**

Long before I knew what calculus was, I sensed there was something special about it. My dad had spoken about it in reverential tones. He hadn't been able to go to college, being a child of the Depression, but somewhere along the line, maybe during his time in the South Pacific repairing B-24 bomber engines, he'd gotten a feel for what calculus could do. Imagine a mechanically controlled bank of antiaircraft guns automatically firing at an incoming fighter plane. Calculus, he supposed, could be used to tell the guns where to aim.

Every year about a million American students take calculus. But far fewer really understand what the subject is about or could tell you why they were learning it. It's not their fault. There are so many techniques to master and so many new ideas to absorb that the overall framework is easy to miss.

Calculus is the mathematics of change. It describes everything from the spread of epidemics to the zigs and zags of a well-thrown curveball. The subject is gargantuan—and so are its textbooks. Many exceed a thousand pages and work nicely as doorstops.

But within that bulk you'll find two ideas shining through. All the rest, as Rabbi Hillel said of the golden rule, is just commentary. Those two ideas are the derivative and the integral.

Each dominates its own half of the subject, named, respectively, differential and integral calculus.

Roughly speaking, the derivative tells you how fast something is changing; the integral tells you how much it's accumulating. They were born in separate times and places: integrals in Greece around 250 B.C.; derivatives in England and Germany in the mid-1600s. Yet in a twist straight out of a Dickens novel, they've turned out to be blood relatives — though it took almost two millennia for anyone to see the family resemblance.

The next chapter will explore that astonishing connection, as well as the meaning of integrals. But first, to lay the groundwork, let's look at derivatives.

Derivatives are all around us, even if we don't recognize them as such. For example, the slope of a ramp is a derivative. Like all derivatives, it measures a rate of change — in this case, how far you're going up or down for every step you take. A steep ramp has a large derivative. A wheelchair-accessible ramp, with its gentle gradient, has a small derivative.

Every field has its own version of a derivative. Whether it goes by marginal return or growth rate or velocity or slope, a derivative by any other name still smells as sweet. Unfortunately, many students seem to come away from calculus with a much narrower interpretation, regarding the derivative as synonymous with the slope of a curve.

Their confusion is understandable. It's caused by our reliance on graphs to express quantitative relationships. By plotting y versus x to visualize how one variable affects another, all scientists translate their problems into the common language of mathematics. The rate of change that really concerns them — a viral growth rate, a jet's velocity, or whatever — then gets converted into something much more abstract but easier to picture: a slope on a graph.

Like slopes, derivatives can be positive, negative, or zero, indicating whether something is rising, falling, or leveling off. Consider Michael Jordan flying through the air before making one of his thunderous dunks.

Just after liftoff, his vertical velocity (the rate at which his elevation changes in time and, thus, another derivative) is positive, because he's going up. His elevation is increasing. On the way down, this derivative is negative. And at the highest point of his jump, where he seems to hang in the air, his elevation is momentarily unchanging and his derivative is zero. In that sense he truly *is* hanging.

There's a more general principle at work here—things always change slowest at the top or the bottom. It's especially noticeable here in Ithaca. During the darkest depths of winter, the days are not just unmercifully short; they barely improve from one to the next. Whereas once spring starts popping, the days lengthen rapidly. All of this makes sense. Change is most sluggish at the extremes precisely because the derivative is zero there. Things stand still, momentarily.

This zero-derivative property of peaks and troughs underlies some of the most practical applications of calculus. It allows us to use derivatives to figure out where a function reaches its maximum or minimum, an issue that arises whenever we're looking for the best or cheapest or fastest way to do something.

My high-school calculus teacher, Mr. Joffray, had a knack for making such "max-min" questions come alive. One day he came bounding into class and began telling us about his hike through a snow-covered field. The wind had apparently blown a lot of snow across part of the field, blanketing it heavily and forcing him to walk much more slowly there, while the rest of the field was clear, allowing him to stride through it easily. In a situation like that, he wondered what path a hiker should take to get from point A to point B as quickly as possible.

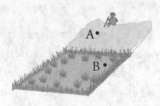

One thought would be to trudge straight across the deep snow, to cut down on the slowest part of the hike. The downside, though, is the rest of the trip will take longer than it would otherwise.

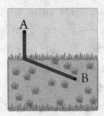

Another strategy is to head straight from A to B. That's certainly the shortest distance, but it does cost extra time in the most arduous part of the trip.

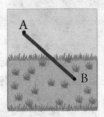

With differential calculus you can find the best path. It's a certain specific compromise between the two paths considered above.

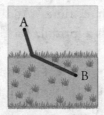

The analysis involves four main steps. (For those who'd like to see the details, references are given in the notes on page 278.)

First, notice that the total time of travel—which is what we're trying to minimize—depends on where the hiker emerges from the snow. He could choose to emerge anywhere, so let's consider all his possible exit points as a variable. Each of these locations can be characterized succinctly by specifying a single number: the distance x where the hiker emerges from the snow.

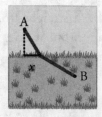

(Implicitly, the travel time also depends on the locations of A and B and on the hiker's speeds in both parts of the field, but all those parameters are given. The only thing under the hiker's control is x.)

Second, given a choice of x and the known locations of the starting point A and the destination B, we can calculate how much time the hiker spends walking through the fast and slow parts of the field. For each leg of the trip, this calculation requires the Pythagorean theorem and the old algebra mantra "distance equals rate times time." Adding the times for both legs together then yields a formula for the total travel time, T, as a function of x.

Third, we graph T versus x. The bottom of the curve is the point we're seeking—it corresponds to the least time of travel and hence the fastest trip.

Fourth, to find this lowest point, we invoke the zero-derivative principle mentioned above. We calculate the derivative of T, set it equal to zero, and solve for x.

These four steps require a command of geometry, algebra, and various derivative formulas from calculus—skills equivalent to fluency in a foreign language and, therefore, stumbling blocks for many students.

But the final answer is worth the struggle. It reveals that the fastest path obeys a relationship known as Snell's law. What's spooky is that nature obeys it too.

Snell's law describes how light rays bend when they pass from air into water, as they do when the sun shines into a swimming pool. Light moves more slowly in water, much like the hiker in the snow, and it bends accordingly to minimize its travel time. Similarly, light bends when it travels from air into glass or plastic, as when it refracts through your eyeglass lenses.

The eerie point is that light behaves as if it were considering all possible paths and then taking the best one. Nature—cue the theme from *The Twilight Zone*—somehow knows calculus.

MATHEMATICAL SIGNS AND symbols are often cryptic, but the best of them offer visual clues to their own meaning. The symbols for zero, one, and infinity aptly resemble an empty hole, a single mark, and an endless loop: 0, 1, ∞. And the equal sign, =, is formed by two parallel lines because, as its originator, Welsh mathematician Robert Recorde, wrote in 1557, "no two things can be more equal."

In calculus the most recognizable icon is the integral sign:

$$\int$$

Its graceful lines are evocative of a musical clef or a violin's f-hole — a fitting coincidence, given that some of the most enchanting harmonies in mathematics are expressed by integrals. But the real reason that the mathematician Gottfried Leibniz chose this symbol is much less poetic. It's simply a long-necked S, for "summation."

As for what's being summed, that depends on context. In astronomy, the gravitational pull of the sun on the Earth is described by an integral. It represents the collective effect of all the minuscule forces generated by each solar atom at their varying distances from the Earth. In oncology, the growing mass of a solid tumor can be modeled by an integral. So can the cumulative amount of drug administered during the course of a chemotherapy regimen.

To understand why sums like these require integral calculus and not the ordinary kind of addition we learned in grade school, let's consider what challenges we'd face if we actually tried to calculate the sun's gravitational pull on the Earth. The first difficulty is that the sun is not a point . . . and neither is the Earth. Both of them are gigantic balls made up of stupendous numbers of atoms. Every atom in the sun exerts a gravitational tug on every atom in the Earth. Of course, since atoms are tiny, their mutual attractions are almost infinitesimally small, yet because there are almost infinitely many of them, in aggregate they can still amount to something. Somehow we have to add them all up.

But there's a second and more serious difficulty: Those pulls are different for different pairs of atoms. Some are stronger than others. Why? Because the strength of gravity *changes* with distance — the closer two things are, the more strongly they attract. The atoms on the far sides of the sun and the Earth feel the least attraction; those on the near sides feel the strongest; and those in between feel forces of middling strength. Integral calculus is needed to sum all those changing forces. Amazingly, it can be done — at least in the idealized limit where we treat the Earth and the sun as solid spheres composed of *infinitely* many points of continuous matter, each exerting an infinitesimal attraction on the others. As in all of calculus: infinity and limits to the rescue!

Historically, integrals arose first in geometry, in connection with the problem of finding the areas of curved shapes. As we saw in chapter 16, the area of a circle can be viewed as the sum of many thin pie slices. In the limit of infinitely many slices, each of which is infinitesimally thin, those slices could then be cunningly rearranged into a rectangle whose area was much easier to find. That was a typical use of integrals. They're all about taking something complicated and slicing and dicing it to make it easier to add up.

In a 3-D generalization of this method, Archimedes (and before him, Eudoxus, around 400 B.C.) calculated the volumes of various solid shapes by reimagining them as stacks of many wafers or disks, like a salami sliced thin. By computing the changing volumes of the varying slices and then ingeniously integrating them — adding them back together — they were able to deduce the volume of the original whole.

Today we still ask budding mathematicians and scientists to sharpen their skills at integration by applying them to these classic geometry problems. They're some of the hardest exercises we assign, and a lot of students hate them, but there's no surer way to hone the facility with integrals needed for advanced work in every quantitative discipline from physics to finance.

One such mind-bender concerns the volume of the solid common to two identical cylinders crossing at right angles, like stovepipes in a kitchen.

It takes an unusual gift of imagination to visualize this three-dimensional shape. So there's no shame in admitting defeat and looking for a way to make it more palpable. To do so, you can resort to a trick my high-school calculus teacher used — take a tin can and cut the top off with metal shears to form a cylindrical coring tool. Then core a large Idaho potato or a piece of Styrofoam from two mutually perpendicular directions. Inspect the resulting shape at your leisure.

Computer graphics now make it possible to visualize this shape more easily.

Remarkably, it has square cross-sections, even though it was created from round cylinders.

It's a stack of infinitely many layers, each a wafer-thin square, tapering from a large square in the middle to progressively smaller ones and finally to single points at the top and bottom.

Still, picturing the shape is merely the first step. What remains is to determine its volume, by tallying the volumes of all the separate slices. Archimedes managed to do this, but only by virtue of his astounding ingenuity. He used a mechanical method based on levers and centers of gravity, in effect weighing the shape in his mind by balancing it against others he already understood. The downside of his approach, besides the prohibitive brilliance it required, was that it applied to only a handful of shapes.

Conceptual roadblocks like this stumped the world's finest mathematicians for the next nineteen centuries . . . until the mid-1600s, when James Gregory, Isaac Barrow, Isaac Newton, and Gottfried Leibniz established what's now known as the fundamental theorem of calculus. It forged a powerful link between the two types of change being studied in calculus: the cumulative change represented by integrals, and the local rate of change represented by derivatives (the subject of chapter 17). By exposing this connection, the fundamental theorem greatly expanded the universe of integrals that could be solved, and it reduced their calculation to grunt work. Nowadays computers can be programmed to use it—and so can students. With its help, even the stovepipe problem that was once a world-class challenge becomes an exercise within common reach. (For the details of Archimedes's approach as well as the modern one, consult the references in the notes on pages 279–280.)

Before calculus and the fundamental theorem came along, only the simplest kinds of net change could be analyzed. When something changes *steadily*, at a constant rate, algebra works beautifully. This is the domain of "distance equals rate times time." For example, a car moving at an unchanging speed of 60 miles per hour will surely travel 60 miles in the first hour, and 120 miles by the end of the second hour.

But what about change that proceeds at a *changing* rate?

Such changing change is all around us—in the accelerating descent of a penny dropped from a tall building, in the ebb and flow of the tides, in the elliptical orbits of the planets, in the circadian rhythms within us. Only calculus can cope with the cumulative effects of changes as non-uniform as these.

For nearly two millennia after Archimedes, just one method existed for predicting the net effect of changing change: add up the varying slices, bit by bit. You were supposed to treat the rate of change as constant within each slice, then invoke the analog of "distance equals rate times time" to inch forward to the end of that slice, and repeat until all the slices were dealt with. Most of the time it couldn't be done. The infinite sums were too hard.

The fundamental theorem enabled a lot of these problems to be solved—not all of them, but many more than before. It often gave a shortcut for solving integrals, at least for the elementary functions (sums and products of powers, exponentials, logarithms, and trig functions) that describe so many of the phenomena in the natural world.

Here's an analogy that I hope will shed some light on what the fundamental theorem says and why it's so helpful. (My colleague Charlie Peskin at New York University suggested it.) Imagine a staircase. The total change in height from the top to the bottom is the sum of the rises of all the steps in between. That's true even if some of the steps rise more than others and no matter how many steps there are.

The fundamental theorem of calculus says something similar for functions—if you integrate the derivative of a function from one point to another, you get the net change in the function between the two points. In this analogy, the function is like the elevation of each step compared to ground level. The rises of individual steps are like the derivative. Integrating the derivative is like summing the rises. And the two points are the top and the bottom.

Why is this so helpful? Suppose you're given a huge list of numbers to sum, as occurs whenever you're calculating an integral by slices. If you can somehow manage to find the corresponding staircase—in other words, if you can find an elevation function for which those numbers are the rises—then computing the integral is a snap. It's just the top minus the bottom.

That's the great speedup made possible by the fundamental theorem. And it's why we torture all beginning calculus students with months of learning how to find elevation functions, technically called antiderivatives or indefinite integrals. This advance allowed mathematicians to forecast events in a changing world with much greater precision than had ever been possible.

From this perspective, the lasting legacy of integral calculus is a Veg-O-Matic view of the universe. Newton and his successors discovered that nature itself unfolds in slices. Virtually all the laws of physics found in the past 300 years turned out to have this character, whether they describe the motions of particles or the flow of heat, electricity, air, or water. Together with the governing laws, the conditions in each slice of time or space determine what will happen in adjacent slices.

The implications were profound. For the first time in history, rational prediction became possible . . . not just one slice at a time but, with the help of the fundamental theorem, by leaps and bounds.

So we're long overdue to update our slogan for integrals —from "It slices, it dices" to "Recalculating. A better route is available."

A FEW NUMBERS ARE SUCH CELEBRITIES that they go by single-letter stage names, something not even Madonna or Prince can match. The most famous is π, the number formerly known as $3.14159\ldots$

Close behind is i, the it-number of algebra, the imaginary number so radical it changed what it meant to be a number. Next on the A list?

Say hello to *e*. Nicknamed for its breakout role in exponential growth, *e* is now the Zelig of advanced mathematics. It pops up everywhere, peeking out from the corners of the stage, teasing us by its presence in incongruous places. For example, along with the insights it offers about chain reactions and population booms, *e* has a thing or two to say about how many people you should date before settling down.

But before we get to that, what is *e*, exactly? Its numerical value is $2.71828\ldots$ but that's not terribly enlightening. I could tell you that *e* equals the limiting number approached by the sum

$$1 + \frac{1}{1} + \frac{1}{1 \times 2} + \frac{1}{1 \times 2 \times 3} + \frac{1}{1 \times 2 \times 3 \times 4} + \cdots$$

as we take more and more terms. But that's not particularly helpful either. Instead, let's look at e in action.

Imagine that you've deposited $1,000 in a savings account at a bank that pays an incredibly generous interest rate of 100 percent, compounded annually. A year later, your account would be worth $2,000—the initial deposit of $1,000 plus the 100 percent interest on it, equal to another $1,000.

Knowing a sucker when you see one, you ask the bank for even more favorable terms: How would they feel about compounding the interest semiannually, meaning that they'd be paying only 50 percent interest for the first six months, followed by another 50 percent for the second six months? You'd clearly do better than before—since you'd gain interest on the interest—but how much better?

The answer is that your initial $1,000 would grow by a factor of 1.50 over the first half of the year, and then by another factor of 1.50 over the second half. And since 1.50 times 1.50 is 2.25, your money would amount to a cool $2,250 after one year, substantially more than the $2,000 you got from the original deal.

What if you pushed even harder and persuaded the bank to divide the year into more and more periods—daily, by the second, or even by the nanosecond? Would you make a small fortune?

To make the numbers come out nicely, here's the result for a year divided into 100 equal periods, after each of which you'd be paid 1 percent interest (the 100 percent annual rate divided evenly into 100 installments): your money would grow by a factor of 1.01 raised to the 100th power, and that comes out to be about 2.70481. In other words, instead of $2,000 or $2,250, you'd have $2,704.81.

Finally, the ultimate: if the interest was compounded *infi-*

nitely often—this is called continuous compounding—your total after one year would be bigger still, but not by much: $2,718.28. The exact answer is $1,000 times *e*, where *e* is defined as the limiting number arising from this process:

$$e = \lim_{n \to \infty} \left(1 + \frac{1}{n}\right)^n = 2.71828\ldots$$

This is a quintessential calculus argument. As we discussed in the last few chapters when we calculated the area of a circle or pondered the sun's gravitational pull on the Earth, what distinguishes calculus from the earlier parts of math is its willingness to confront—and harness—the awesome power of infinity. Whether we're looking at limits, derivatives, or integrals, we always have to sidle up to infinity in one way or another.

In the limiting process that led to *e* above, we imagined slicing a year into more and more compounding periods, windows of time that became thinner and thinner, ultimately approaching what can only be described as infinitely many, infinitesimally thin windows. (This might sound paradoxical, but it's really no worse than treating a circle as the limit of a regular polygon with more and more sides, each of which gets shorter and shorter.) The fascinating thing is that the more often the interest is compounded, the less your money grows during each period. Yet it still amounts to something substantial after a year, because it's been multiplied over so many periods!

This is a clue to the ubiquity of *e*. It often arises when something changes through the cumulative effect of many tiny events.

Consider a lump of uranium undergoing radioactive decay. Moment by moment, every atom has a certain small chance of disintegrating. Whether and when each one does is completely unpredictable, and each such event has an infinitesimal effect on the whole. Nevertheless, in ensemble these trillions of events produce a smooth, predictable, exponentially decaying level of radioactivity.

Or think about the world's population, which grows approximately exponentially. All around the world, children are being born at random times and places, while other people are dying, also at random times and places. Each event has a minuscule impact, percentagewise, on the world's overall population — yet in aggregate that population grows exponentially at a very predictable rate.

Another recipe for e combines randomness with enormous numbers of choices. Let me give you two examples inspired by everyday life, though in highly stylized form.

Imagine there's a very popular new movie showing at the local theater. It's a romantic comedy, and hundreds of couples (many more than the theater can accommodate) are lined up at the box office, desperate to get in. Once a lucky couple get their tickets, they scramble inside and choose two seats right next to each other. To keep things simple, let's suppose they choose these seats at random, wherever there's room. In other words, they don't care whether they sit close to the screen or far away, on the aisle or in the middle of a row. As long as they're together, side by side, they're happy.

Also, let's assume no couple will ever slide over to make room for another. Once a couple sits down, that's it. No courtesy whatsoever. Knowing this, the box office stops selling tickets as soon as there are only single seats left. Otherwise brawls would ensue.

At first, when the theater is pretty empty, there's no problem. Every couple can find two adjacent seats. But after a while, the only seats left are singles — solitary, uninhabitable dead spaces that a couple can't use. In real life, people often create these buffers deliberately, either for their coats or to avoid sharing an armrest with a repulsive stranger. In this model, however, these dead spaces just happen by chance.

The question is: When there's no room left for any more couples, what fraction of the theater's seats are unoccupied?

The answer, in the case of a theater with many seats per row, turns out to approach

$$\frac{1}{e^2} = 0.135 \ldots$$

so about 13.5 percent of the seats go to waste.

Although the details of the calculation are too intricate to present here, it's easy to see that 13.5 percent is in the right ballpark by comparing it with two extreme cases. If all couples sat next to each other, packed in with perfect efficiency like sardines, there'd be no wasted seats.

However, if they'd positioned themselves as *inefficiently* as possible, always with an empty seat between them (and leaving
an empty aisle seat on one end or the other of each row, as in
the diagram below), one-third of the seats would be wasted,
because every couple uses three seats: two for themselves, and
one for the dead space.

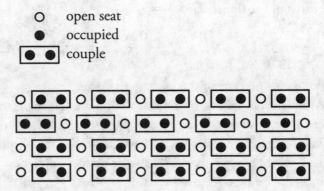

Guessing that the random case should lie somewhere between
perfect efficiency and perfect inefficiency, and taking the aver-
age of 0 and $\frac{1}{3}$, we'd expect that about $\frac{1}{6}$, or 16.7 percent, of
the seats would be wasted, not far from the exact answer of
13.5 percent.

Here the large number of choices came about because of all
the ways that couples could be arranged in a huge theater. Our
final example is also about arranging couples, except now in
time, not space. What I'm referring to is the vexing problem of
how many people you should date before choosing a mate. The
real-life version of this problem is too hard for math, so let's
consider a simplified model. Despite its unrealistic assumptions, it still captures some of the heartbreaking uncertainties
of romance.

Let's suppose you know how many potential mates you're going to meet during your lifetime. (The actual number is not important as long as it's known ahead of time and it's not too small.)

Also assume you could rank these people unambiguously if you could see them all at once. The tragedy, of course, is that you can't. You meet them one at a time, in random order. So you can never be sure if Dreamboat—who'd rank number 1 on your list—is just around the corner, or whether you've already met and parted.

And the way this game works is, once you let someone go, he or she is gone. No second chances.

Finally, assume you don't want to settle. If you end up with Second Best, or anyone else who, in retrospect, wouldn't have made the top of your list, you'll consider your love life a failure.

Is there any hope of choosing your one true love? If so, what can you do to give yourself the best odds?

A good strategy, though not the best one, is to divide your dating life into two equal halves. In the first half, you're just playing the field; in the second, you're ready to get serious, and you're going to grab the first person you meet who's better than everyone else you've dated so far.

With this strategy, there's at least a 25 percent chance of snagging Dreamboat. Here's why: You have a 50-50 chance of meeting Dreamboat in the second half of your dating life, your "get serious" phase, and a 50-50 chance of meeting Second Best in the first half, while you're playing the field. If both of those things happen—and there is a 25 percent chance that they will—then you'll end up with your one true love.

That's because Second Best raised the bar so high. No one you meet after you're ready to get serious will tempt you except Dreamboat. So even though you can't be sure at the time that

Dreamboat is, in fact, The One, that's who he or she will turn out to be, since no one else can clear the bar set by Second Best.

The optimal strategy, however, is to stop playing the field a little sooner, after only $1/e$, or about 37 percent, of your potential dating lifetime. That gives you a $1/e$ chance of ending up with Dreamboat.

As long as Dreamboat isn't playing the e game too.

"IN THE SPRING," wrote Tennyson, "a young man's fancy lightly turns to thoughts of love." Alas, his would-be partner has thoughts of her own — and the interplay between them can lead to the tumultuous ups and downs that make new love so thrilling, and so painful. To explain these swings, many love-lorn souls have sought answers in drink; others have turned to poetry. We'll consult calculus.

The analysis below is offered tongue-in-cheek, but it touches on a serious point: While the laws of the heart may elude us forever, the laws of inanimate things are now well understood. They take the form of differential equations, which describe how interlinked variables change from moment to moment, depending on their current values. As for what such equations have to do with romance — well, at the very least they might shed a little light on why, in the words of another poet, "the course of true love never did run smooth."

To illustrate the approach, suppose Romeo is in love with Juliet but that, in our version of the story, Juliet is a fickle lover. The more Romeo loves her, the more she wants to run away and hide. But when he takes the hint and backs off, she begins to find him strangely attractive. He, however, tends to mirror her: he warms up when she loves him and cools down when she hates him.

What happens to our star-crossed lovers? How does their love ebb and flow over time? That's where calculus comes in. By writing equations that summarize how Romeo and Juliet respond to each other's affections and then solving those equations with calculus, we can predict the course of their affair. The resulting forecast for this couple is, tragically, a never-ending cycle of love and hate. At least they manage to achieve simultaneous love a quarter of the time.

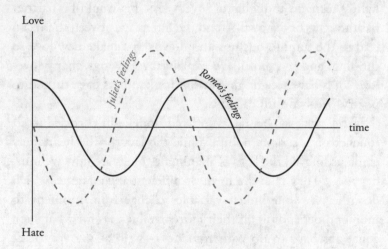

To reach this conclusion, I've assumed that Romeo's behavior can be modeled by the differential equation

$$\frac{dR}{dt} = aJ,$$

which describes how his love (represented by R) changes in the next instant (represented by dt). According to this equation, the amount of change (dR) is just a multiple (a) of Juliet's current love (J) for him. This reflects what we already know—that Romeo's love goes up when Juliet loves him—but it assumes something much stronger. It says that Romeo's love increases in direct linear proportion to how much Juliet loves him. This assumption of linearity is not emotionally realistic, but it makes the equations much easier to solve.

Juliet's behavior, by contrast, can be modeled by the equation

$$\frac{dJ}{dt} = -bR.$$

The negative sign in front of the constant b reflects her tendency to cool off when Romeo is hot for her.

The only remaining thing we need to know is how the lovers felt about each other initially (R and J at time $t = 0$). Then everything about their affair is predetermined. We can use a computer to inch R and J forward, instant by instant, changing their values as prescribed by the differential equations. Actually, with the help of the fundamental theorem of calculus, we can do much better than that. Because the model is so simple, we don't have to trudge forward one moment at a time. Calculus yields a pair of comprehensive formulas that tell us how much Romeo and Juliet will love (or hate) each other at *any* future time.

The differential equations above should be recognizable to students of physics: Romeo and Juliet behave like simple harmonic oscillators. So the model predicts that $R(t)$ and $J(t)$—the functions that describe the time course of their relationship—will be sine waves, each waxing and waning but peaking at different times.

The model can be made more realistic in various ways. For instance, Romeo might react to his own feelings as well as to Juliet's. He might be the type of guy who is so worried about throwing himself at her that he slows himself down as his love for her grows. Or he might be the other type, one who loves feeling in love so much that he loves her all the more for it.

Add to those possibilities the two ways Romeo could react to Juliet's affections—either increasing or decreasing his own—and you see that there are four personality types, each corresponding to a different romantic style. My students and those in Peter Christopher's class at Worcester Polytechnic Institute have suggested such descriptive names as Hermit and Malevolent Misanthrope for the particular kind of Romeo who damps down his own love and also recoils from Juliet's. Whereas the sort of Romeo who gets pumped by his own ardor but turned off by Juliet's has been called Narcissistic Nerd, Better Latent Than Never, and a Flirting Fink. (Feel free to come up with your own names for these two types and the other two possibilities.)

Although these examples are whimsical, the kinds of equations that arise in them are profound. They represent the most powerful tool humanity has ever created for making sense of the material world. Sir Isaac Newton used differential equations to solve the ancient mystery of planetary motion. In so doing, he unified the earthly and celestial spheres, showing that the same laws of motion applied to both.

In the nearly 350 years since Newton, mankind has come to realize that the laws of physics are always expressed in the language of differential equations. This is true for the equations governing the flow of heat, air, and water; for the laws of electricity and magnetism; even for the unfamiliar and often counterintuitive atomic realm, where quantum mechanics reigns.

In all cases, the business of theoretical physics boils down to finding the right differential equations and solving them. When Newton discovered this key to the secrets of the universe, he felt it was so precious that he published it only as an anagram in Latin. Loosely translated, it reads: "It is useful to solve differential equations."

The silly idea that love affairs might likewise be described by differential equations occurred to me when I was in love for the first time, trying to understand my girlfriend's baffling behavior. It was a summer romance at the end of my sophomore year in college. I was a lot like the first Romeo above, and she was even more like the first Juliet. The cycling of our relationship drove me crazy until I realized that we were both acting mechanically, following simple rules of push and pull. But by the end of the summer my equations started to break down, and I was more mystified than ever. As it turned out, there was an important variable that I'd left out of the equations—her old boyfriend wanted her back.

In mathematics we call this a three-body problem. It's notoriously intractable, especially in the astronomical context where it first arose. After Newton solved the differential equations for the two-body problem (thus explaining why the planets move in elliptical orbits around the sun), he turned his attention to the three-body problem for the sun, Earth, and moon. He couldn't solve it, and neither could anyone else. It later turned out that the three-body problem contained the seeds of chaos, rendering its behavior unpredictable in the long run.

Newton knew nothing about chaotic dynamics, but according to his friend Edmund Halley, he complained that the three-body problem had "made his head ache, and *kept him awake so often, that he would think of it no more.*"

I'm with you there, Sir Isaac.

Mr. DICURCIO WAS my mentor in high school. He was disagreeable and demanding, with nerdy black-rimmed glasses and a penchant for sarcasm, so his charms were easy to miss. But I found his passion for physics irresistible.

One day I mentioned to him that I was reading a biography of Einstein. The book said that as a college student, Einstein had been dazzled by something called Maxwell's equations for electricity and magnetism, and I said I couldn't wait until I knew enough math to learn what they were.

This being a boarding school, we were eating dinner together at a big table with several other students, his wife, and his two daughters, and Mr. DiCurcio was serving mashed potatoes. At the mention of Maxwell's equations, he dropped the serving spoon, grabbed a paper napkin, and began writing lines of cryptic symbols—dots and crosses, upside-down triangles, Es and Bs with arrows over them—and suddenly he seemed to be speaking in tongues: "The curl of a curl is grad div minus del squared . . ."

That abracadabra he was mumbling? I realize now he was speaking in vector calculus, the branch of math that describes the invisible fields all around us. Think of the magnetic field that twists a compass needle northward, or the gravitational field that pulls your chair to the floor, or the microwave field that nukes your dinner.

The greatest achievements of vector calculus lie in that twilight realm where math meets reality. Indeed, the story of James Clerk Maxwell and his equations offers one of the eeriest instances of the unreasonable effectiveness of mathematics. Somehow, by shuffling a few symbols, Maxwell discovered what light is.

To give a sense of what Maxwell accomplished and, more generally, what vector calculus is about, let's begin with the word "vector." It comes from the Latin root *vehere*, "to carry," which also gives us words like "vehicle" and "conveyor belt." To an epidemiologist, a vector is the carrier of a pathogen, like the mosquito that conveys malaria to your bloodstream. To a mathematician, a vector (at least in its simplest form) is a step that carries you from one place to another.

Think about one of those diagrams for aspiring ballroom dancers covered with arrows indicating how to move the right foot, then the left foot, as when doing the rumba:

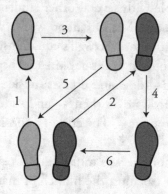

These arrows are vectors. They show two kinds of information: a direction (which way to move that foot) and a magnitude (how far to move it). All vectors do that same double duty.

Vectors can be added and subtracted, just like numbers, except their directionality makes things a little trickier. Still, the right way to add vectors becomes clear if you think of them as dance instructions. For example, what do you get when you take one step east followed by one step north? A vector that points northeast, naturally.

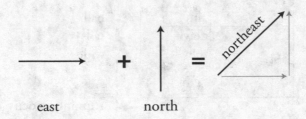

Remarkably, velocities and forces work the same way — they too add just like dance steps. This should be familiar to any tennis player who's ever tried to imitate Pete Sampras and hit a forehand down the line while sprinting at full speed toward the sideline. If you naively aim your shot where you want it to go, it will sail wide because you forgot to take your own running into account. The ball's velocity relative to the court is the sum of *two* vectors: the ball's velocity relative to you (a vector pointing down the line, as intended), and your velocity relative to the court (a vector pointing sideways, since that's the direction you're running). To hit the ball where you want it to go, you have to aim slightly crosscourt, to compensate for your sideways motion.

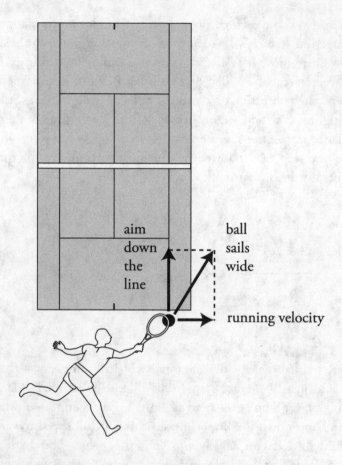

Beyond such vector algebra lies vector calculus, the kind of math Mr. DiCurcio was using. Calculus, you'll recall, is the mathematics of change. And so whatever vector calculus is, it must involve vectors that change, either from moment to moment or from place to place. In the latter case, one speaks of a "vector field."

A classic example is the force field around a magnet. To visualize it, put a magnet on a piece of paper and sprinkle iron

filings everywhere. Each filing acts like a little compass needle—it aligns with the direction of local "north," determined
by the magnetic field at that point. Viewed in aggregate, these
filings reveal a spectacular pattern of magnetic-field lines leading from one pole of the magnet to the other.

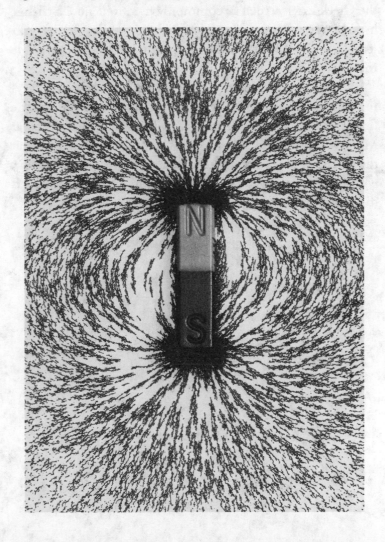

The direction and magnitude of the vectors in a magnetic field vary from point to point. As in all of calculus, the key tool for quantifying such changes is the derivative. In vector calculus the derivative operator goes by the name of del, which has a folksy southern ring to it, though it actually alludes to the Greek letter Δ (delta), commonly used to denote a change in some variable. As a reminder of that kinship, "del" is written like this: ∇. (That was the mysterious upside-down triangle Mr. DiCurcio kept writing on the napkin.)

It turns out there are two different but equally natural ways to take the derivative of a vector field by applying del to it. The first gives what's known as the field's divergence (the "div" that Mr. DiCurcio muttered). To get an intuitive feeling for what the divergence measures, take a look at the vector field below, which shows how water would flow from a source on the left to a sink on the right.

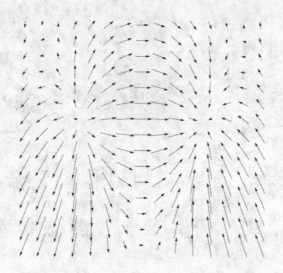

For this example, instead of using iron filings to track the vector field, imagine lots of tiny corks or bits of leaves floating on the water surface. We're going to use them as probes. Their motion will tell us how the water is moving at each point. Specifically, imagine what would happen if we put a small circle of corks around the source. Obviously, the corks would spread apart and the circle would expand, because water flows away from a source. It *diverges* there. And the stronger the divergence, the faster the area of our cork-circle would grow. That's what the divergence of a vector field measures: how fast the area of a small circle of corks grows.

The image below shows the numerical value of the divergence at each point in the field we've just been looking at, coded by shades of gray. Lighter shades show points where the flow has a positive divergence. Darker shades show places of negative divergence, meaning that the flow would *compress* a tiny cork-circle centered there.

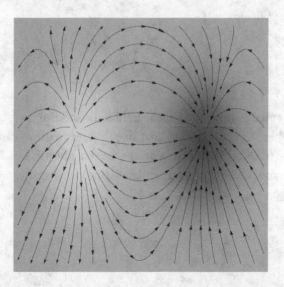

The other kind of derivative measures the curl of a vector field. Roughly speaking, it indicates how strongly the field is swirling about a given point. (Think of the weather maps you've seen on the local news showing the rotating wind patterns around hurricanes or tropical storms.) In the vector field below, regions that look like hurricanes have a large curl.

By embellishing the vector field with shading, we can now show where the curl is most positive (lightest regions) and most negative (darkest regions). Notice that this also tells us whether the flow is spinning counterclockwise or clockwise.

The curl is extremely informative for scientists working in fluid mechanics and aerodynamics. A few years ago my colleague Jane Wang used a computer to simulate the pattern of airflow around a dragonfly as it hovered in place. By calculating the curl, she found that when a dragonfly flaps its wings, it creates pairs of counter-rotating vortices that act like little tornadoes beneath its wings, producing enough lift to keep the insect aloft. In this way, vector calculus is helping to explain how dragonflies, bumblebees, and hummingbirds can fly — something that had long been a mystery to conventional fixed-wing aerodynamics.

With the notions of divergence and curl in hand, we're now ready to revisit Maxwell's equations. They express four fundamental laws: one for the divergence of the electric field, another for its curl, and two more of the same type but now for the magnetic field. The divergence equations relate the electric and magnetic fields to their sources, the charged particles and currents that produce them in the first place. The curl equations describe how the electric and magnetic fields interact and change over time. In so doing, these equations express a beautiful symmetry: they link one field's rate of change in *time* to the *other* field's rate of change in *space*, as quantified by its curl.

Using mathematical maneuvers equivalent to vector calculus—which wasn't known in his day—Maxwell then extracted the logical consequences of those four equations. His symbol shuffling led him to the conclusion that electric and magnetic fields could propagate as a wave, somewhat like a ripple on a pond, except that these two fields were more like symbiotic organisms. Each sustained the other. The electric field's undulations re-created the magnetic field, which in turn re-created the electric field, and so on, with each pulling the other forward, something neither could do on its own.

That was the first breakthrough—the theoretical prediction of electromagnetic waves. But the real stunner came next. When Maxwell calculated the speed of these hypothetical waves, using known properties of electricity and magnetism, his equations told him that they traveled at about 193,000 miles per second—the same rate as the speed of light measured by the French physicist Hippolyte Fizeau a decade earlier!

How I wish I could have witnessed the moment when a human being first understood the true nature of light. By identifying it with an electromagnetic wave, Maxwell unified three ancient and seemingly unrelated phenomena: electricity, magnetism, and light. Although experimenters like Faraday and Ampère had previously found key pieces of this puzzle, it was only Maxwell, armed with his mathematics, who put them all together.

Today we are awash in Maxwell's once-hypothetical waves: Radio. Television. Cell phones. Wi-Fi. These are the legacy of his conjuring with symbols.

Part Five **DATA**

The New Normal

STATISTICS HAS SUDDENLY become cool and trendy. Thanks to the emergence of the Internet, e-commerce, social networks, the Human Genome Project, and digital culture in general, the world is now teeming with data. Marketers scrutinize our habits and tastes. Intelligence agencies collect data on our whereabouts, e-mails, and phone calls. Sports statisticians crunch the numbers to decide which players to trade, whom to draft, and whether to go for it on fourth down with two yards to go. Everybody wants to connect the dots, to find the needle of meaning in the haystack of data.

So it's not surprising that students are being advised accordingly. "Learn some statistics," exhorted Greg Mankiw, an economist at Harvard, in a 2010 column in the *New York Times*. "High school mathematics curriculums spend too much time on traditional topics like Euclidean geometry and trigonometry. For a typical person, these are useful intellectual exercises but have little applicability to daily life. Students would be better served by learning more about probability and statistics." David Brooks put it more bluntly. In a column about what college courses everyone should take to be properly educated, he

wrote, "Take statistics. Sorry, but you'll find later in life that it's
handy to know what a standard deviation is."

Yes, and even handier to know what a distribution is. That's
the first idea I'd like to focus on here, because it embodies one
of the central lessons of statistics — things that seem hopelessly
random and unpredictable when viewed in isolation often turn
out to be lawful and predictable when viewed in aggregate.

You may have seen a demonstration of this principle at
a science museum (if not, you can find videos online). The
standard exhibit involves a setup called a Galton board, which
looks a bit like a pinball machine except it has no flippers and
its bumpers consist of a regular array of evenly spaced pegs arranged in rows.

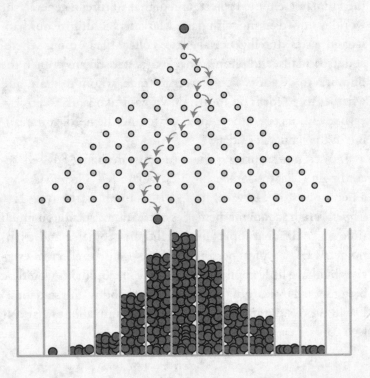

The demo begins when hundreds of balls are poured into the top of the Galton board. As they rain down, they randomly bounce off the pegs, sometimes to the left, sometimes to the right, and ultimately distribute themselves in the evenly spaced bins at the bottom. The height of the stacked balls in each bin shows how likely it was for a ball to land there. Most balls end up somewhere near the middle, with slightly fewer balls flanking them on either side, and fewer still far off in the tails at either end. Overall, the pattern is utterly predictable: it always forms a bell-shaped distribution — even though it's impossible to predict where any given ball will end up.

How does individual randomness turn into collective regularity? Easy — the odds demand it. The middle bin is likely to be the most populated spot because most balls will make about the same number of leftward and rightward bounces before they reach the bottom. So they'll probably end up somewhere near the middle. The only balls that make it far out to either extreme, way off in the tails of the distribution where the outliers live, are those that just happen to bounce in the same direction almost every time they hit a peg. That's very unlikely. And that's why there are so few balls out there.

Just as the ultimate location of each ball is determined by the sum of many chance events, lots of phenomena in this world are the net result of many tiny flukes, so they too are governed by a bell-shaped curve. Insurance companies bank on this. They know, with great accuracy, how many of their customers will die each year. They just don't know who the unlucky ones will be.

Or consider how tall you are. Your height depends on countless little accidents of genetics, biochemistry, nutrition, and environment. Consequently, it's plausible that when viewed in aggregate, the heights of adult men and women should fall on a bell curve.

In a blog post titled "The big lies people tell in online dating," the statistically minded dating service OkCupid recently graphed how tall their members are—or rather, how tall they *say* they are—and found that the heights reported by both sexes follow bell curves, as expected. What's surprising, however, is that both distributions are shifted about two inches to the right of where they should be.

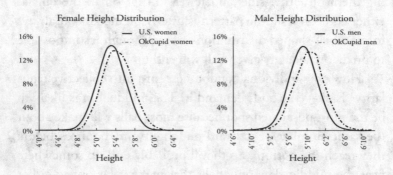

So either the people who join OkCupid are unusually tall, or they exaggerate their heights by a couple of inches when describing themselves online.

An idealized version of these bell curves is what mathematicians call the normal distribution. It's one of the most important concepts in statistics. Part of its appeal is theoretical. The normal distribution can be proven to arise whenever a large number of mildly random effects of similar size, all acting independently, are added together. And many things are like that.

But not everything. That's the second point I'd like to stress. The normal distribution is not nearly as ubiquitous as it once seemed. For about 100 years now, and especially during the past few decades, statisticians and scientists have noticed that plenty of phenomena deviate from this pattern yet still manage to follow a pattern of their own. Curiously, these types of distributions are barely mentioned in the elementary statistics textbooks, and when they are, they're usually trotted out as pathological specimens. It's outrageous. For, as I'll try to explain, much of modern life makes a lot more sense when you understand these distributions. They're the new normal.

Take the distribution of city sizes in the United States. Instead of clustering around some intermediate value in a bell-shaped fashion, the vast majority of towns and cities are tiny and therefore huddle together on the left of the graph.

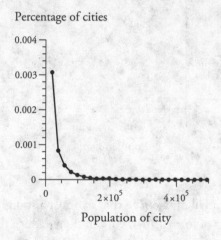

And the larger the population of a city, the more uncommon a city of that size is. So when viewed in the aggregate, the distribution looks more like an L-curve than a bell curve.

There's nothing surprising about this. Everybody knows that big cities are rarer than small ones. What's less obvious, though, is that city sizes nevertheless follow a beautifully simple distribution . . . as long as you look at them through logarithmic lenses.

In other words, suppose we regard the size differential between a pair of cities to be the same if their populations differ by the same *factor*, rather than by the same absolute number of people (much like two pitches an octave apart always differ by a constant factor of double the frequency). And suppose we do likewise on the vertical axis.

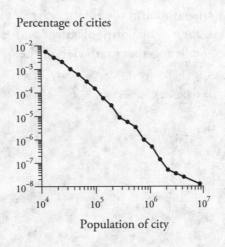

Then the data fall on a curve that's almost a straight line. From the properties of logarithms, we can then deduce that the original L-curve was a power law, a function of the form

$$y = \frac{C}{x^a},$$

where x is a city's size, y is how many cities have that size, C is a constant, and the exponent a (the power in the power law) is the negative of the straight line's slope.

Power-law distributions have counterintuitive properties from the standpoint of conventional statistics. For example, unlike normal distributions', their modes, medians, and means do not agree because of the skewed, asymmetrical shapes of their L-curves. President Bush made use of this property when he stated that his 2003 tax cuts had saved families an average of $1,586 each. Though that is technically correct, he was conveniently referring to the *mean* rebate, a figure that averaged in the whopping rebates of hundreds of thousands of dollars received by the richest 0.1 percent of the population. The tail on the far right of the income distribution is known to follow a power law, and in situations like this, the mean is a misleading statistic to use because it's far from typical. Most families, in fact, got less than $650 back. The median was a lot less than the mean.

This example highlights the most crucial feature of power-law distributions. Their tails are heavy (also known as fat or long), at least compared to the puny little wisp of a tail on the normal distribution. So extremely large outliers, though still rare, are much more common for these distributions than they would be for normal bell curves.

On October 19, 1987, now known as Black Monday, the Dow Jones industrial average dropped by 22 percent in a single day. Compared to the usual level of volatility in the stock market, this was a drop of more than twenty standard deviations. Such an event is all but impossible according to traditional bell-curve statistics; its probability is less than one in 100,000,000,000,000,000,000,000,000,000,000,000,000,000, 000,000,000,000 (that's 10 raised to the 50th power). Yet it happened . . . because fluctuations in stock prices don't follow normal distributions. They're better described by heavy-tailed distributions.

So are earthquakes, wildfires, and floods, which complicates the task of risk management for insurance industries. The same mathematical pattern holds for the numbers of deaths caused by wars and terrorist attacks, and even for more benign things like word frequencies in novels and the number of sexual partners people have.

Though the adjectives used to describe their prominent tails weren't originally meant to be flattering, such distributions have come to wear them with pride. Fat, heavy, and long? Yeah, that's right. Now who's normal?

HAVE YOU EVER had that anxiety dream where you suddenly realize you have to take the final exam in some course you've never attended? For professors, it works the other way around—you dream you're giving a lecture in a course you know nothing about.

That's what it's like for me whenever I teach probability theory. It was never part of my own education, so having to lecture about it now is scary and fun, in an amusement-park-thrill-house sort of way.

Perhaps the most pulse-quickening topic of all is conditional probability—the probability that some event A happens, given (or conditional upon) the occurrence of some other event B. It's a slippery concept, easily conflated with the probability of B given A. They're not the same, but you have to concentrate to see why. For example, consider the following word problem.

Before going on vacation for a week, you ask your spacy friend to water your ailing plant. Without water, the plant has a 90 percent chance of dying. Even with proper watering, it has a 20 percent chance of dying. And the probability that your friend will forget to water it is 30 percent. (a) What's the chance that your plant will survive the week? (b) If it's dead when you return, what's the chance that your friend forgot to water it? (c) If your friend forgot to water it, what's the chance it'll be dead when you return? Although they sound alike, (b) and (c) are not the same. In fact, the problem tells us that the answer to (c) is 90 percent. But how do you combine all the probabilities to get the answer to (b)? Or (a)?

Naturally, the first few semesters I taught this topic, I stuck to the book, inching along, playing it safe. But gradually I began to notice something. A few of my students would avoid using Bayes's theorem, the labyrinthine formula I was teaching them. Instead they would solve the problems by an equivalent method that seemed easier.

What these resourceful students kept discovering, year after year, was a better way to think about conditional probability. Their way comports with human intuition instead of confounding it. The trick is to think in terms of natural frequencies—simple counts of events—rather than in more abstract notions of percentages, odds, or probabilities. As soon as you make this mental shift, the fog lifts.

This is the central lesson of *Calculated Risks*, a fascinating book by Gerd Gigerenzer, a cognitive psychologist at the Max Planck Institute for Human Development in Berlin. In a series of studies about medical and legal issues ranging from AIDS counseling to the interpretation of DNA fingerprinting, Gigerenzer explores how people miscalculate risk and uncertainty. But rather than scold or bemoan human frailty, he tells us how to do better—how to avoid clouded thinking by recasting conditional-probability problems in terms of natural frequencies, much as my students did.

In one study, Gigerenzer and his colleagues asked doctors in Germany and the United States to estimate the probability that a woman who has a positive mammogram actually has breast cancer even though she's in a low-risk group: forty to fifty years old, with no symptoms or family history of breast cancer. To make the question specific, the doctors were told to assume the following statistics—couched in terms of percentages and probabilities—about the prevalence of breast cancer among women in this cohort and about the mammogram's sensitivity and rate of false positives:

> The probability that one of these women has breast cancer is 0.8 percent. If a woman has breast cancer, the probability is 90 percent that she will have a positive mammogram. If a woman does not have breast cancer, the probability is 7 percent that she will still have a positive mammogram. Imagine a woman who has a positive mammogram. What is the probability that she actually has breast cancer?

Gigerenzer describes the reaction of the first doctor he tested, a department chief at a university teaching hospital with more than thirty years of professional experience:

> [He] was visibly nervous while trying to figure out what he would tell the woman. After mulling the numbers over, he finally estimated the woman's probability of having breast cancer, given that she has a positive mammogram, to be 90 percent. Nervously, he added, "Oh, what nonsense. I can't do this. You should test my daughter; she is studying medicine." He knew that his estimate was wrong, but he did not know how to reason better. Despite the fact that he had spent 10 minutes wringing his mind for an answer, he could

not figure out how to draw a sound inference from the probabilities.

Gigerenzer asked twenty-four other German doctors the same question, and their estimates whipsawed from 1 percent to 90 percent. Eight of them thought the chances were 10 percent or less; eight others said 90 percent; and the remaining eight guessed somewhere between 50 and 80 percent. Imagine how upsetting it would be as a patient to hear such divergent opinions.

As for the American doctors, ninety-five out of a hundred estimated the woman's probability of having breast cancer to be somewhere around 75 percent.

The right answer is 9 percent.

How can it be so low? Gigerenzer's point is that the analysis becomes almost transparent if we translate the original information from percentages and probabilities into natural frequencies:

> Eight out of every 1,000 women have breast cancer. Of these 8 women with breast cancer, 7 will have a positive mammogram. Of the remaining 992 women who don't have breast cancer, some 70 will still have a positive mammogram. Imagine a sample of women who have positive mammograms in screening. How many of these women actually have breast cancer?

Since a total of 7 + 70 = 77 women have positive mammograms, and only 7 of them truly have breast cancer, the probability of a woman's having breast cancer given a positive mammogram is 7 out of 77, which is 1 in 11, or about 9 percent.

Notice two simplifications in the calculation above. First,

we rounded off decimals to whole numbers. That happened in a few places, like when we said, "Of these 8 women with breast cancer, 7 will have a positive mammogram." Really we should have said 90 percent of 8 women, or 7.2 women, will have a positive mammogram. So we sacrificed a little precision for a lot of clarity.

Second, we assumed that everything happens exactly as frequently as its probability suggests. For instance, since the probability of breast cancer is 0.8 percent, exactly 8 women out of 1,000 in our hypothetical sample were assumed to have it. In reality, this wouldn't necessarily be true. Events don't have to follow their probabilities; a coin flipped 1,000 times doesn't always come up heads 500 times. But pretending that it does gives the right answer in problems like this.

Admittedly the logic is a little shaky — that's why the textbooks look down their noses at this approach, compared to the more rigorous but hard-to-use Bayes's theorem — but the gains in clarity are justification enough. When Gigerenzer tested another set of twenty-four doctors, this time using natural frequencies, nearly all of them got the correct answer, or close to it.

Although reformulating the data in terms of natural frequencies is a huge help, conditional-probability problems can still be perplexing for other reasons. It's easy to ask the wrong question or to calculate a probability that's correct but misleading.

Both the prosecution and the defense were guilty of this in the O.J. Simpson trial of 1994–95. Each of them asked the court to consider the wrong conditional probability.

The prosecution spent the first ten days of the trial introducing evidence that O.J. had a history of violence toward his ex-wife Nicole Brown. He had allegedly battered her, thrown her against walls, and groped her in public, telling onlookers, "This belongs to me." But what did any of this have to do with a murder trial? The prosecution's argument was that a pattern of spousal abuse reflected a motive to kill. As one of the prosecutors put it, "A slap is a prelude to homicide."

Alan Dershowitz countered for the defense, arguing that even if the allegations of domestic violence were true, they were irrelevant and should therefore be inadmissible. He later wrote, "We knew we could prove, if we had to, that an infinitesimal percentage—certainly fewer than 1 of 2,500—of men who slap or beat their domestic partners go on to murder them."

In effect, both sides were asking the court to consider the probability that a man murdered his ex-wife, given that he previously battered her. But as the statistician I. J. Good pointed out, that's not the right number to look at.

The real question is: What's the probability that a man murdered his ex-wife, given that he previously battered her *and she was murdered by someone*? That conditional probability turns out to be very far from 1 in 2,500.

To see why, imagine a sample of 100,000 battered women. Granting Dershowitz's number of 1 in 2,500, we expect about 40 of these women to be murdered by their abusers in a given year (since 100,000 divided by 2,500 equals 40). We also expect 3 more of these women, on average, to be killed by someone else (this estimate is based on statistics reported by the FBI for women murdered in 1992; see the notes for further details). So out of the 43 murder victims altogether, 40 of them were killed by their batterers. In other words, the batterer was the murderer about 93 percent of the time.

Don't confuse this number with the probability that O.J. did it. That probability would depend on a lot of other evidence, pro and con, such as the defense's claim that the police framed O.J., and the prosecution's claim that the killer and O.J. shared the same style of shoes, gloves, and DNA.

The probability that any of this changed your mind about the verdict? Zero.

Untangling the Web

IN A TIME long ago, in the dark days before Google, searching the Web was an exercise in frustration. The sites suggested by the older search engines were too often irrelevant, while the ones you really wanted were either buried way down in the list of results or missing altogether.

Algorithms based on link analysis solved the problem with an insight as paradoxical as a Zen koan: A Web search should return the best pages. And what, grasshopper, makes a page good? A page is good if other good pages link to it.

That sounds like circular reasoning. It is . . . which is why it's so deep. By grappling with this circle and turning it to advantage, link analysis yields a jujitsu solution to searching the Web.

The approach builds on ideas from linear algebra, the study of vectors and matrices. Whether you want to detect patterns in large data sets or perform gigantic computations involving millions of variables, linear algebra has the tools you need. Along with underpinning Google's PageRank algorithm, it has helped scientists classify human faces, analyze the voting patterns of Supreme Court justices, and win the million-dollar Netflix Prize (awarded to the person or team who could improve by

more than 10 percent Netflix's system for recommending mov-
ies to its customers).

For a case study of linear algebra in action, let's look at how
PageRank works. And to bring out its essence with a minimum
of fuss, let's imagine a toy Web that has just three pages, all
connected like this:

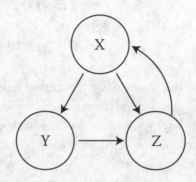

The arrows indicate that page X contains a link to page Y, but
Y does not return the favor. Instead, Y links to Z. Meanwhile X
and Z link to each other in a frenzy of digital back-scratching.

In this little Web, which page is the most important, and
which is the least? You might think there's not enough information to say because nothing is known about the pages' content.
But that's old-school thinking. Worrying about content turned
out to be an impractical way to rank webpages. Computers
weren't good at it, and human judges couldn't keep up with the
deluge of thousands of pages added each day.

The approach taken by Larry Page and Sergey Brin, the
grad students who cofounded Google, was to let webpages
rank themselves by voting with their feet—or, rather, with
their links. In the example above, pages X and Y both link to Z,

which makes Z the only page with two incoming links. So it's the most popular page in the universe. That should count for something. However, if those links come from pages of dubious quality, that should count against them. Popularity means nothing on its own. What matters is having links from *good* pages.

Which brings us back to the riddle of the circle: A page is good if good pages link to it, but who decides which pages are good in the first place?

The network does. And here's how. (Actually, I'm skipping some details; see the notes on pages 292–293 for a more complete story.)

Google's algorithm assigns a fractional score between 0 and 1 to each page. That score is called its PageRank; it measures how important that page is relative to the others by computing the proportion of time that a hypothetical Web surfer would spend there. Whenever there is more than one outgoing link to choose from, the surfer selects one at random, with equal probability. Under this interpretation, pages are regarded as more important if they're visited more frequently (by this idealized surfer, not by actual Web traffic).

And because the PageRanks are defined as proportions, they have to add up to 1 when summed over the whole network. This conservation law suggests another, perhaps more palpable, way to visualize PageRank. Picture it as a fluid, a watery substance that flows through the network, draining away from bad pages and pooling at good ones. The algorithm seeks to determine how this fluid distributes itself across the network in the long run.

The answer emerges from a clever iterative process. The algorithm starts with a guess, then updates all the PageRanks by apportioning the fluid in equal shares to the outgoing links, and it keeps doing that in a series of rounds until everything settles down and all the pages get their rightful shares.

Initially the algorithm takes an egalitarian stance. It gives every page an equal portion of PageRank. Since there are three pages in the example we're considering, each page begins with a score of 1/3.

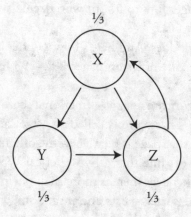

Initial values of PageRank

Next, these scores are updated to better reflect each page's true importance. The rule is that each page takes its PageRank from the last round and parcels it out equally to all the pages it links to. Thus, after one round, the updated value of X would still equal 1/3, because that's how much PageRank it receives from Z, the only page that links to it. But Y's score drops to a measly 1/6, since it gets only half of X's PageRank from the previous round. The other half goes to Z, which makes Z the big winner at this stage, since along with the 1/6 it receives from X, it also gets the full 1/3 from Y, for a total of 1/2. So after one round, the PageRank values are those shown below:

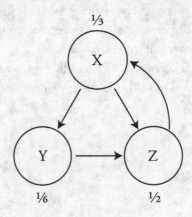

Values of PageRank
after one update

In the rounds to come, the update rule stays the same. If we write (x, y, z) for the current scores of pages X, Y, and Z, then the updated scores will be

$$x' = z$$
$$y' = \tfrac{1}{2} x$$
$$z' = \tfrac{1}{2} x + y$$

where the prime symbol in the superscript signifies that an update has occurred. This kind of iterative calculation is easy to do in a spreadsheet (or even by hand, for a network as small as the one we're studying).

After ten iterations, one finds that the numbers don't change much from one round to the next. By then, X has a 40.6 percent share of the total PageRank, Y has 19.8 percent, and Z has 39.6 percent. Those numbers look suspiciously close to 40 percent, 20 percent, and 40 percent, suggesting that the algorithm is converging to those values.

In fact, that's correct. Those limiting values are what Google's algorithm would define as *the* PageRanks for the network.

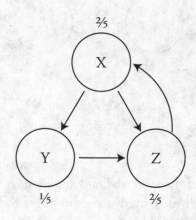

Limiting values of PageRank

The implication is that X and Z are equally important pages, even though Z has twice as many links coming in. That makes sense: X is just as important as Z because it gets the full endorsement of Z but reciprocates with only half its own endorsement. The other half it sends to Y. This also explains why Y fares only half as well as X and Z.

Remarkably, these scores can be obtained directly, without

going through the iteration. Just think about the conditions that define the steady state. If nothing changes after an update is performed, we must have $x' = x$, $y' = y$, and $z' = z$. So replace the primed variables in the update equations with their unprimed counterparts. Then we get

$$x = z$$
$$y = \tfrac{1}{2} x$$
$$z = \tfrac{1}{2} x + y$$

and this system of equations can be solved simultaneously to obtain $x = 2y = z$. Finally, since these scores must sum to 1, we conclude $x = 2/5$, $y = 1/5$, and $z = 2/5$, in agreement with the percentages found above.

Let's step back for a moment to look at how all this fits into the larger context of linear algebra. The steady-state equations above, as well as the earlier update equations with the primes in them, are typical examples of linear equations. They're called linear because they're related to lines. The variables x, y, z in them appear to the first power only, just as they do in the familiar equation for a straight line, $y = mx + b$, a staple of high-school algebra courses.

Linear equations, as opposed to those containing nonlinear terms like x^2 or yz or $\sin x$, are comparatively easy to solve. The challenge comes when there are enormous numbers of variables involved, as there are in the real Web. One of the central tasks of linear algebra, therefore, is the development of faster and faster algorithms for solving such huge sets of equations. Even slight improvements have ramifications for everything from airline scheduling to image compression.

But the greatest triumph of linear algebra, from the standpoint of real-world impact, is surely its solution to the Zen riddle of ranking webpages. "A page is good insofar as good pages link to it." Translated into symbols, that criterion becomes the PageRank equations.

Google got where it is today by solving the same equations as we did here—just with a few billion more variables . . . and profits to match.

Part Six FRONTIERS

ACCORDING TO A memorable song from the 1960s, one is the loneliest number, and two can be as bad as one. Maybe so, but the prime numbers have it pretty rough too.

Paolo Giordano explains why in his best-selling novel *The Solitude of Prime Numbers*. It's the melancholy love story of two misfits, two primes, named Mattia and Alice, both scarred by childhood tragedies that left them virtually incapable of connecting with other people, yet who sense in each other a kindred damaged spirit. Giordano writes,

> Prime numbers are divisible only by 1 and by themselves. They hold their place in the infinite series of natural numbers, squashed, like all numbers, between two others, but one step further than the rest. They are suspicious, solitary numbers, which is why Mattia thought they were wonderful. Sometimes he thought that they had ended up in that sequence by mistake, that they'd been trapped, like pearls strung on a necklace. Other times he suspected that they too would have preferred to be like all the others, just ordinary

numbers, but for some reason they couldn't do it. [. . .]

In his first year at university, Mattia had learned that, among prime numbers, there are some that are even more special. Mathematicians call them twin primes: pairs of prime numbers that are close to each other, almost neighbors, but between them there is always an even number that prevents them from truly touching. Numbers like 11 and 13, like 17 and 19, 41 and 43. If you have the patience to go on counting, you discover that these pairs gradually become rarer. You encounter increasingly isolated primes, lost in that silent, measured space made only of ciphers, and you develop a distressing presentiment that the pairs encountered up until that point were accidental, that solitude is the true destiny. Then, just when you're about to surrender, when you no longer have the desire to go on counting, you come across another pair of twins, clutching each other tightly. There is a common conviction among mathematicians that however far you go, there will always be another two, even if no one can say where exactly, until they are discovered.

Mattia thought that he and Alice were like that, twin primes, alone and lost, close but not close enough to really touch each other.

Here I'd like to explore some of the beautiful ideas in the passage above, particularly as they relate to the solitude of prime numbers and twin primes. These issues are central to number theory, the subject that concerns itself with the study of whole numbers and their properties and that is often described as the purest part of mathematics.

Before we ascend to where the air is thin, let me dispense with a question that often occurs to practical-minded people: Is number theory good for anything? Yes. Almost in spite of itself, number theory provides the basis for the encryption algorithms used millions of times each day to secure credit card transactions over the Internet and to encode military-strength secret communications. Those algorithms rely on the difficulty of decomposing an enormous number into its prime factors.

But that's not why mathematicians are obsessed with prime numbers. The real reason is that they're fundamental. They're the atoms of arithmetic. Just as the Greek origin of the word "atom" suggests, the primes are "a-tomic," meaning "uncuttable, indivisible." And just as everything is composed of atoms, every number is composed of primes. For example, 60 equals $2 \times 2 \times 3 \times 5$. We say that 60 is a composite number with prime factors of 2 (counted twice), 3, and 5.

And what about 1? Is it prime? No, it isn't, and when you understand why it isn't, you'll begin to appreciate why 1 truly is the loneliest number—even lonelier than the primes.

It doesn't deserve to be left out. Given that 1 is divisible only by 1 and itself, it really should be considered prime, and for many years it was. But modern mathematicians have decided to exclude it, solely for convenience. If 1 were allowed in, it would mess up a theorem that we'd like to be true. In other words, we've rigged the definition of prime numbers to give us the theorem we want.

The desired theorem says that any number can be factored into primes in a *unique* way. But if 1 were considered prime, the uniqueness of prime factorization would fail. For example, 6 would equal 2×3, but it would also equal $1 \times 2 \times 3$ and $1 \times 1 \times 2 \times 3$ and so on, and these would all have to be accepted as different prime factorizations. Silly, of course, but that's what we'd be stuck with if 1 were allowed in.

This sordid little tale is instructive; it pulls back the curtain on how math is done sometimes. The naive view is that we make our definitions, set them in stone, then deduce whatever theorems happen to follow from them. Not so. That would be much too passive. We're in charge and can alter the definitions as we please — especially if a slight tweak leads to a tidier theorem, as it does here.

Now that 1 has been thrown under the bus, let's look at everyone else, the full-fledged prime numbers. The main thing to know about them is how mysterious they are, how alien and inscrutable. No one has ever found an exact formula for the primes. Unlike real atoms, they don't follow any simple pattern, nothing akin to the periodic table of the elements.

You can already see the warning signs in the first ten primes: 2, 3, 5, 7, 11, 13, 17, 19, 23, 29. Right off the bat, things start badly with 2. It's a freak, a misfit among misfits — the only prime with the embarrassment of being an even number. No wonder "it's the loneliest number since the number one" (as the song says).

Apart from 2, the rest of the primes are all odd . . . but still quirky. Look at the gaps between them. Sometimes they're two spaces apart (like 5 and 7), sometimes four (13 and 17), and sometimes six (23 and 29).

To further underscore how disorderly the primes are, compare them to their straight-arrow cousins the odd numbers: 1, 3, 5, 7, 9, 11, 13, . . . The gaps between odd numbers are always consistent: two spaces, steady as a drumbeat. So they obey a simple formula: the nth odd number is $2n - 1$. The primes, by contrast, march to their own drummer, to a rhythm no one else can perceive.

Given the irregularities in the spacing of the primes, numbers theorists have resorted to looking at them statistically, as members of an ensemble, rather than dwelling on their idiosyncrasies. Specifically, let's ask how they're distributed among the ordinary whole numbers. How many primes are less than or equal to 10? Or 100? Or an arbitrary number N? This construction is a direct parallel to the statistical concept of a cumulative distribution.

So imagine counting the prime numbers by walking among them, like an anthropologist taking a census. Picture them standing there on the x-axis. You start at the number 1 and begin walking to the right, tallying primes as you go. Your running total looks like this:

Number of primes ≤ x

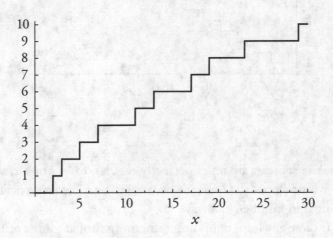

The values on the y-axis show how many primes you've counted by the time you reach a given location, x. For all x's less than 2, the graph of y remains flat at 0, since no primes have been counted yet. The first prime appears at $x = 2$. So the graph jumps up there. (Got one!) Then it remains flat until $x = 3$, after which it jumps up another step. The alternating jumps and plateaus form a strange, irregular staircase. Mathematicians call it the counting function for the primes.

Contrast this image with its counterpart for the odd numbers.

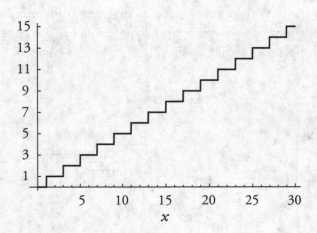

Number of odd numbers ≤ x

Now the staircase becomes perfectly regular, following a trend line whose slope is 1/2. That's because the gap between neighboring odd numbers is always 2.

Is there any hope of finding something similar for the prime numbers despite their erratic character? Miraculously, yes. The key is to focus on the trend, not the details of the stair steps. If we zoom out, a curve begins to emerge from the clutter. Here's the graph of the counting function for all primes up to 100.

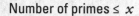

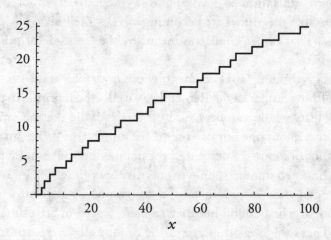

The steps are less distracting now. The curve looks even smoother if we count all the primes out to a billion:

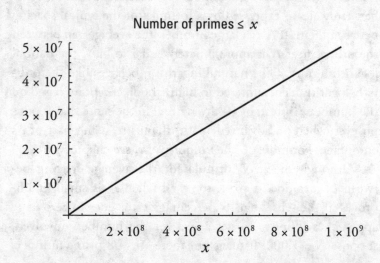

First impressions to the contrary, this curve is *not* actually a straight line. It droops down slightly as it climbs. Its droopiness means that the primes are becoming *rarer*. More isolated. More alone. That's what Giordano meant by the "solitude of prime numbers."

This thinning out becomes obvious if we look at the census data from another angle. Remember we tallied ten primes in the first thirty whole numbers. So near the beginning of the number line, about one out of every three numbers is prime, making them a robust 33 percent of the population. But among the first hundred numbers, only twenty-five are prime. Their ranks have dwindled to one in four, a worrisome 25 percent. And among the first billion numbers, a mere 5 percent are prime.

That's the bleak message of the droopy curve. The primes are a dying breed. They never die out completely — we've known since Euclid they go on forever — but they fade into near oblivion.

By finding functions that approximate the droopy curve, number theorists have quantified how desolate the prime numbers truly are, as expressed by a formula for the typical spacing between them. If N is a large number, the average gap between the primes near N is approximately equal to $\ln N$, the natural logarithm of N. (The natural logarithm behaves like the ordinary logarithm encountered in high school, except it's based on the number e instead of 10. It's natural in the sense that it pops up everywhere in advanced math, thanks to being part of e's entourage. For more on the ubiquity of e, see chapter 19.)

Although the $\ln N$ formula for the average spacing between primes doesn't work too well when N is small, it improves in the sense that its percentage error goes to zero as N approaches infinity. To get a feel for the numbers involved, suppose $N = 1,000$. It turns out there are 168 prime numbers less than 1,000, so the average gap between them in this part of the number line is 1,000/68, or about 5.9. For comparison, the formula predicts an average gap of ln(1,000) ≈ 6.9, which is too high by about 17 percent. But when we go much farther out, say to N = 1,000,000,000, the actual and predicted gaps become 19.7 and 20.7, respectively, an overestimate of only about 5 percent.

The validity of the ln N formula as N tends to infinity is now known as the prime number theorem. It was first noticed (but not published) by Carl Friedrich Gauss in 1792 when he was fifteen years old. (See what a kid can do when not distracted by an Xbox?)

As for this chapter's other teens, Mattia and Alice, I hope you can appreciate how poignant it is that twin primes apparently continue to exist in the farthest reaches of the number line, "in that silent, measured space made only of ciphers." The odds are stacked against them. According to the prime number theorem, any particular prime near N has no right to expect a potential mate much closer than ln N away, a gulf much larger than 2 when N is large.

And yet some couples do beat the odds. Computers have found twin primes at unbelievably remote parts of the number line. The largest known pair consists of two numbers with 100,355 decimal digits each, snuggling in the darkness.

The twin prime conjecture says couples like this will turn up forever.

But as for finding another prime couple nearby for a friendly game of doubles? Good luck.

MY WIFE AND I have different sleeping styles—and our mattress shows it. She hoards the pillows, thrashes around all night long, and barely dents the mattress, while I lie on my back, mummy-like, molding a cavernous depression into my side of the bed.

Bed manufacturers recommend flipping your mattress periodically, probably with people like me in mind. But what's the best system? How exactly are you supposed to flip it to get the most even wear out of it?

Brian Hayes explores this problem in the title essay of his book *Group Theory in the Bedroom*. Double-entendres aside, the "group" under discussion here is a collection of mathematical actions—all the possible ways you could flip or rotate the mattress so that it still fits neatly on the bed frame.

By looking into mattress math in some detail, I hope to give you a feeling for group theory more generally. It's one of the most versatile parts of mathematics. It underlies everything from the choreography of square dancing and the fundamental laws of particle physics to the mosaics of the Alhambra and their chaotic counterparts, like this image:

As these examples suggest, group theory bridges the arts and sciences. It addresses something the two cultures share — an abiding fascination with symmetry. Yet because it encompasses such a wide range of phenomena, group theory is necessarily abstract. It distills symmetry to its essence.

Normally we think of symmetry as a property of a shape. But group theorists focus more on what you can *do* to a shape — specifically, all the ways you can change it while keeping something else about it the same. More precisely, they look for all the transformations that leave a shape unchanged, given certain constraints. These transformations are called the symmetries of the shape. Taken together, they form a group, a collection of transformations whose relationships define the shape's most basic architecture.

In the case of a mattress, the transformations alter its orientation in space (that's what changes) while maintaining its rigidity (that's the constraint). And after the transformation is complete, the mattress has to fit snugly on the rectangular bed frame (that's what stays the same). With these rules in place, let's see what transformations qualify for membership in this exclusive little group. It turns out there are only four of them.

The first is the do-nothing transformation, a lazy but popular choice that leaves the mattress untouched. It certainly satisfies all the rules, but it's not much help in prolonging the life of your mattress. Still, it's very important to include in the group. It plays the same role for group theory that 0 does for addition of numbers, or that 1 does for multiplication. Mathematicians call it the identity element, so I'll denote it by the symbol I.

Next come the three genuine ways to flip a mattress. To distinguish among them, it helps to label the corners of the mattress by numbering them like so:

The first kind of flip is depicted near the beginning of this chapter. The handsome gentleman in striped pajamas is trying to turn the mattress from side to side by rotating it 180 degrees around its long axis, in a move I'll call *H*, for "horizontal flip."

Horizontal Flip

A more reckless way to turn over the mattress is a vertical flip, *V*. This maneuver swaps its head and foot. You stand the mattress upright the long way, so that it almost reaches the ceiling, and then topple it end over end. The net effect, besides the enormous thud, is that the mattress rotates 180 degrees about its lateral axis, shown below.

Vertical Flip

The final possibility is to spin the mattress half a turn while keeping it flat on the bed.

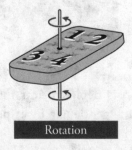

Rotation

Unlike the *H* and *V* flips, this rotation, *R*, keeps the top surface on top.

That difference shows up when we look at the mattress—now imagined to be translucent—from above and inspect the numbers at the corners after each of the possible transformations. The horizontal flip turns the numerals into their mirror images. It also permutes them so that 1 and 2 trade places, as do 3 and 4.

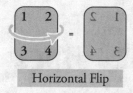

Horizontal Flip

The vertical flip permutes the numbers in a different way and stands them on their heads, in addition to mirroring them.

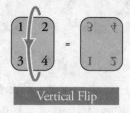

Vertical Flip

The rotation, however, doesn't generate any mirror images. It merely turns the numbers upside down, this time exchanging 1 for 4 and 2 for 3.

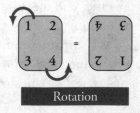

Rotation

These details are not the main point. What matters is how the transformations relate to one another. Their patterns of interaction encode the symmetry of the mattress.

To reveal those patterns with a minimum of effort, it helps to draw the following diagram. (Images like this abound in a terrific book called *Visual Group Theory*, by Nathan Carter. It's one of the best introductions to group theory—or to any branch of higher math—I've ever read.)

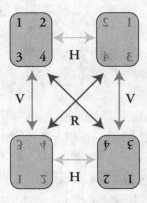

The four possible states of the mattress are shown at the corners of the diagram. The upper left state is the starting point. The arrows indicate the moves that take the mattress from one state to another.

For example, the arrow pointing from the upper left to the lower right depicts the action of the rotation *R*. The arrow also has an arrowhead on the other end, because if you do *R* twice, it's tantamount to doing nothing.

That shouldn't come as a surprise. It just means that turning the mattress head to foot and then doing that again returns the mattress to its original state. We can summarize this property with the equation $RR = I$, where RR means "do R twice," and I is the do-nothing identity element. The horizontal and vertical flip transformations also undo themselves: $HH = I$ and $VV = I$.

The diagram embodies a wealth of other information. For instance, it shows that the death-defying vertical flip, V, is equivalent to HR, a horizontal flip followed by a rotation — a much safer path to the same result. To check this, begin at the starting state in the upper left. Head due east along H to the next state, and from there go diagonally southwest along R. Because you arrive at the same state you'd reach if you followed V to begin with, the diagram demonstrates that $HR = V$.

Notice, too, that the order of those actions is irrelevant: $HR = RH$, since both roads lead to V. This indifference to order is true for any other pair of actions. You should think of this as a generalization of the commutative law for addition of ordinary numbers, x and y, according to which $x + y = y + x$. But beware: The mattress group is special. Many other groups violate the commutative law. Those fortunate enough to obey it are particularly clean and simple.

Now for the payoff. The diagram shows how to get the most even wear out of a mattress. Any strategy that samples all four states periodically will work. For example, alternating R and H is convenient — and since it bypasses V, it's not too strenuous. To help you remember it, some manufacturers suggest the mnemonic "spin in the spring, flip in the fall."

The mattress group also pops up in some unexpected places, from the symmetry of water molecules to the logic of a pair of electrical switches. That's one of the charms of group theory. It exposes the hidden unity of things that would otherwise seem unrelated . . . like in this anecdote about how the physicist Richard Feynman got a draft deferment.

The army psychiatrist questioning him asked Feynman to put out his hands so he could examine them. Feynman stuck them out, one palm up, the other down. "No, the other way," said the psychiatrist. So Feynman reversed *both* hands, leaving one palm down and the other up.

Feynman wasn't merely playing mind games; he was indulging in a little group-theoretic humor. If we consider all the possible ways he could have held out his hands, along with the various transitions among them, the arrows form the same pattern as the mattress group!

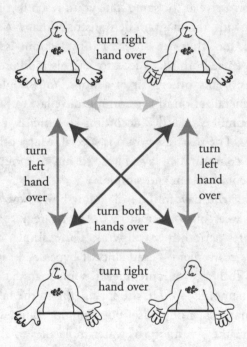

But if all this makes mattresses seem way too complicated, maybe the real lesson here is one you already knew—if something's bothering you, just sleep on it.

Twist and Shout

Our local elementary school invites parents to come and talk to their children's classes. This gives the kids a chance to hear about different jobs and to learn about things they might not be exposed to otherwise.

When my turn came, I showed up at my daughter's first-grade class with a bag full of Möbius strips. The night before, my wife and I had cut long strips of paper and then given each strip a half twist, like this,

before taping the two ends together to make a Möbius strip:

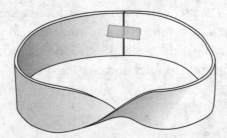

There are fun activities that a six-year-old can do with these shapes that involve nothing more than scissors, crayons, tape, and a little curiosity.

As my wife and I passed out the Möbius strips and art supplies, the teacher asked the class what subject they thought we were doing now. One boy raised his hand and said, "Well, I'm not sure, but I know it's not linguistics."

Of course, the teacher had been expecting an answer of "art," or maybe, more precociously, "math." The best answer, however, would have been "topology." (In Ithaca it's altogether possible that a first-grader could have come up with that. But the topologist's kid happened to be in another class that year.)

So what is topology? It's a vibrant branch of modern math, an offshoot of geometry, but much more loosey-goosey. In topology, two shapes are regarded as the same if you can bend, twist, stretch, or otherwise deform one into the other continuously—that is, without any ripping or puncturing. Unlike the rigid objects of geometry, the objects of topology behave as if they were infinitely elastic, as if they were made of an ideal kind of rubber or Silly Putty.

Topology shines a spotlight on a shape's deepest properties—the properties that remain unchanged after a continuous distortion. For example, a rubber band shaped like a square and another shaped like a circle are topologically indistinguishable. It doesn't matter that a square has four corners and four straight sides. Those properties are irrelevant. A continuous deformation can get rid of them by rounding off the square's corners and bending its sides into circular arcs.

But the one thing such a deformation *can't* get rid of is the intrinsic loopiness of a circle and a square. They're both closed curves. That's their shared topological essence.

Likewise, the essence of a Möbius strip is the peculiar half twist locked into it. That twist endows the shape with its signature features. Most famously, a Möbius strip has only one side and only one edge. In other words, its front and back surfaces are actually the same, and so are its top and bottom edges. (To check this, just run your finger along the middle of the strip or its edge until you return to your starting point.) What happened here is that the half twist hooked the former top and bottom edges of the paper into one big, long continuous curve. Similarly, it fused the front to the back. Once the strip has been taped shut, these features are permanent. You can stretch and twist a Möbius strip all you want, but nothing can change its half-twistedness, its one-sidedness, and its one-edgedness.

By having the first-graders examine the strange properties of Möbius strips that follow from these features, I was hoping to show them how much fun math could be—and also how amazing.

First I asked them each to take a crayon and carefully draw a line all the way around the Möbius strip, right down the middle of the surface. With brows furrowed, they began tracing something like the dashed line on the following page:

After one circuit, many of the students stopped and looked puzzled. Then they began shouting to one another excitedly, because their lines had not closed as they'd expected. The crayon had not come back to the starting point; it was now on the "other" side of the surface. That was surprise number one: you have to go *twice* around a Möbius strip to get back to where you started.

Suddenly one boy began melting down. When he realized his crayon hadn't come back to its starting point, he thought he'd done something wrong. No matter how many times we reassured him that this was supposed to happen, that he was doing a great job, and that he should just go around the strip one more time, it didn't help. It was too late. He was on the floor, wailing, inconsolable.

With some trepidation I asked the class to try the next activity. What did they think would happen if they took their scissors and cut neatly down the midline all the way along the length of the strip?

It will fall apart! It will make two pieces! they guessed. But after they tried it and something incredible happened (the strip remained in one piece but grew twice as long), there were even more squeals of surprise and delight. It was like a magic trick.

After that it was hard to hold the students' attention. They were too busy trying their own experiments, making new kinds of Möbius strips with two or three half twists in them and cutting them lengthwise into halves, thirds, or quarters, producing all sorts of twisted necklaces, chains, and knots, all the while shouting variations of "Hey, look what I found!" But I still can't get over that one little boy I traumatized. And apparently my lesson wasn't the first to have driven a student to tears.

Vi Hart was so frustrated by her boring math courses in high school that she began doodling in class, sketching snakes and trees and infinite chains of shrinking elephants, instead of listening to the teacher droning on. Vi, who calls herself a "full-time recreational mathemusician," has posted some of her doodles on YouTube. They've now been watched hundreds of thousands of times, and in the case of the elephants, more than a million. She, and her videos, are breathtakingly original.

Two of my favorites highlight the freaky properties of Möbius strips through an inventive use of music and stories. In the less baffling of the two, her "Möbius music box" plays a theme from a piece of music she composed, inspired by the Harry Potter books.

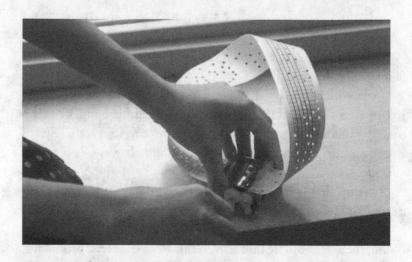

The melody is encoded as a series of holes punched through a tape, which is then fed through a standard music box. Her innovation was to twist the ends of the tape and join them together to form a Möbius strip. By cranking the crank on the music box, Vi feeds the tape through the device and the melody plays in the normal fashion. But about fifty seconds into the video, the loop completes one circuit, and because of the half twist in the Möbius strip, the music box now begins playing what was originally the *back* of the punched tape, upside down. Hence the same melody begins again but now with all the notes *inverted*. High notes become low notes, and low notes become high. They're still played in the same order, but upside down, thanks to the somersaults imposed by the Möbius structure.

For an even more striking example of the topsy-turvy implications of Möbius strips, in "Möbius Story: Wind and Mr. Ug," Vi tells a bittersweet story of unattainable love. A friendly little triangle named Wind, drawn with an erasable marker, unknowingly lives in a flat world made out of clear acetate and shaped like a Möbius strip. She's lonesome but ever hopeful, eager to meet the world's only other inhabitant, a mysterious gent named Mr. Ug who lives one door down. Although she's never met him—he always seems to be out when she stops by his house—she loves the messages he leaves for her and longs to meet him someday.

Spoiler alert: skip the next paragraph if you don't want to learn the story's secret.

Mr. Ug doesn't exist. Wind is Mr. Ug, viewed upside down and on the back of the transparent Möbius strip. Because of the clever way that Vi prints letters and makes the world turn by spinning the acetate, when Wind's name or her house or her messages go once around the Möbius strip, all those things flip over and look like they belong to Mr. Ug.

My explanation doesn't do this video justice. You've simply got to watch it to see the tremendous ingenuity at work here, combining a unique love story with vivid illustrations of the properties of Möbius strips.

Other artists have likewise drawn inspiration from the perplexing features of Möbius strips. Escher used them in his drawings of ants trapped in an eternal loop. Sculptors and stone carvers, like Max Bill and Keizo Ushio, have incorporated Möbius motifs in their massive works.

Perhaps the most monumental of all Möbius structures is that being planned for the National Library of Kazakhstan. Its design, by the Danish architectural firm BIG, calls for spiraling public paths that coil up and then down and in which "like a yurt the wall becomes the roof, which becomes floor, which becomes the wall again."

The properties of Möbius strips offer design advantages to engineers as well. For example, a continuous-loop recording tape in the shape of a Möbius strip doubles the playing time.

The B. F. Goodrich Company patented a Möbius strip conveyor belt, which lasts twice as long as a conventional belt since it wears evenly on "both" sides of its surface (you know what I mean). Other Möbius patents include novel designs for capacitors, abdominal surgical retractors, and self-cleaning filters for dry-cleaning machines.

But perhaps the niftiest application of topology is one that doesn't involve Möbius strips at all. It's a variation on the theme of twists and links, and you might find it helpful next time you have guests over for brunch on a Sunday morning. It's the work of George Hart, Vi's dad. He's a geometer and a sculptor, formerly a computer science professor at Stony Brook University and the chief of content at MoMath, the Museum of Mathematics in New York City. George has devised a way to slice a bagel in half such that the two pieces are locked together like the links of a chain.

The advantage, besides leaving your guests agog, is that it creates more surface area—and hence more room for the cream cheese.

THE MOST FAMILIAR ideas of geometry were inspired by an ancient vision—a vision of the world as flat. From the Pythagorean theorem to parallel lines that never meet, these are eternal truths about an imaginary place, the two-dimensional landscape of plane geometry.

Conceived in India, China, Egypt, and Babylonia more than 2,500 years ago and codified and refined by Euclid and the Greeks, this flat-earth geometry is the main one (and often the only one) being taught in high schools today. But things have changed in the past few millennia.

In an era of globalization, Google Earth, and intercontinental air travel, all of us should try to learn a little about spherical geometry and its modern generalization, differential geometry. The basic ideas here are only about 200 years old. Pioneered by Carl Friedrich Gauss and Bernhard Riemann, differential geometry underpins such imposing intellectual edifices as Einstein's general theory of relativity. At its heart, however, are beautiful concepts that can be grasped by anyone who's ever ridden a bicycle, looked at a globe, or stretched a rubber band. And understanding them will help you make sense of a few curiosities you may have noticed in your travels.

For example, when I was little, my dad used to enjoy quizzing me about geography. Which is farther north, he'd ask, Rome or New York City? Most people would guess New York City, but surprisingly they're at almost the same latitude, with Rome being just a bit farther north. On the usual map of the world (the misleading Mercator projection, where Greenland appears gigantic), it looks like you could go straight from New York to Rome by heading due east.

Yet airline pilots never take that route. They always fly northeast out of New York, hugging the coast of Canada. I used to think they were staying close to land for safety's sake, but that's not the reason. It's simply the most direct route when you take the Earth's curvature into account. The shortest way to get from New York to Rome is to go past Nova Scotia and Newfoundland, head out over the Atlantic, and finally pass south of Ireland and fly across France to arrive in sunny Italy.

This kind of path on the globe is called an arc of a great circle. Like straight lines in ordinary space, great circles on a sphere contain the shortest paths between any two points. They're called great because they're the largest circles you can have on a sphere. Conspicuous examples include the equator and the longitudinal circles that pass through the North and South Poles.

Another property that lines and great circles share is that they're the straightest paths between two points. That might sound strange — *all* paths on a globe are curved, so what do we mean by "straightest"? Well, some paths are more curved than others. The great circles don't do any *additional* curving above and beyond what they're forced to do by following the surface of the sphere.

Here's a way to visualize this. Imagine you're riding a tiny bicycle on the surface of a globe, and you're trying to stay on a certain path. If it's part of a great circle, you can keep the front wheel pointed straight ahead at all times. That's the sense in which great circles are straight. In contrast, if you try to ride along a line of latitude near one of the poles, you'll have to keep the handlebars turned.

Of course, as surfaces go, the plane and the sphere are abnormally simple. The surface of a human body, or a tin can, or a bagel would be more typical — they all have far less symmetry, as well as various kinds of holes and passageways that make them more confusing to navigate. In this more general setting, finding the shortest path between any two points becomes a lot trickier. So rather than delving into technicalities, let's stick to an intuitive approach. This is where rubber bands come in handy.

Specifically, imagine a slippery elastic string that always

contracts as far as it can while remaining on an object's surface. With its help, we can easily determine the shortest path between New York and Rome or, for that matter, between any two points on any surface. Tie the ends of the string to the points of departure and arrival and let the string pull itself tight while it continues clinging to the surface's contours. When the string is as taut as these constraints allow, voilà! It traces the shortest path.

On surfaces just a little more complicated than planes or spheres, something strange and new can happen: *many* locally shortest paths can exist between the same two points. For example, consider the surface of a soup can, with one point lying directly below the other.

Then the shortest path between them is clearly a line segment, as shown above, and our elastic string would find that solution. So what's new here? The cylindrical shape of the can opens up new possibilities for all kinds of contortions. Suppose we require that the string encircles the cylinder once before connecting to the second point. (Constraints like this are imposed on DNA when it wraps around certain proteins in chromosomes.) Now when the string pulls itself taut, it forms a helix, like the curves on old barbershop poles.

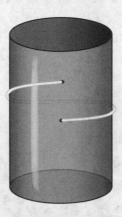

This helical path qualifies as another solution to the shortest-path problem, in the sense that it's the shortest of the candidate paths *nearby*. If you nudge the string a little, it would necessarily get longer and then contract back to the helix. You could say it's the locally shortest path — the regional champion of all those that wrap once around the cylinder. (By the way, this is why the subject is called differential geometry: it studies the effects of small local *differences* on various kinds of shapes, such as the difference in length between the helical path and its neighbors.)

But that's not all. There's another champ that winds around twice, and another that goes around three times, and so on. There are infinitely many locally shortest paths on a cylinder! Of course, none of these helices is the globally shortest path. The straight-line path is shorter than all of them.

Likewise, surfaces with holes and handles permit many locally shortest paths, distinguished by their pattern of weaving around various parts of the surface. The following snapshot from a video by the mathematician Konrad Polthier of the Free University of Berlin illustrates the non-uniqueness of these locally shortest paths, or geodesics, on the surface of an imaginary planet shaped like a figure eight, a surface known in the trade as a two-holed torus:

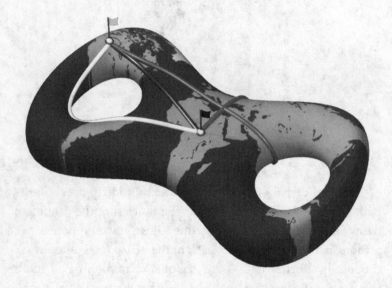

The three geodesics shown here visit very different parts of the planet, thereby executing different loop patterns. But what they all have in common is their superior directness compared to the paths nearby. And just like lines on a plane or great circles on a sphere, these geodesics are the straightest possible curves on the surface. They bend to conform to the surface but don't bend *within* it. To make this clear, Polthier has produced another illuminating video.

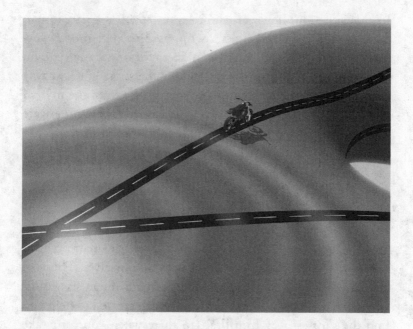

Here, an unmanned motorcycle rides along a geodesic highway on a two-holed torus, following the lay of the land. The remarkable thing is that its handlebars are locked straight ahead; it doesn't need to steer to stay on the road. This underscores the earlier impression that geodesics, like great circles, are the natural generalization of straight lines.

With all these flights of fancy, you may be wondering if geodesics have anything to do with reality. Of course they do. Einstein showed that light beams follow geodesics as they sail through the universe. The famous bending of starlight around the sun, detected in the eclipse observations of 1919, confirmed that light travels on geodesics through curved space-time, with the warping being caused by the sun's gravity.

At a more down-to-earth level, the mathematics of finding shortest paths is critical to the routing of traffic on the Internet.

In this situation, however, the relevant space is a gargantuan maze of addresses and links, as opposed to the smooth surfaces considered above, and the mathematical issues have to do with the speed of algorithms—what's the most efficient way to find the shortest path through a network? Given the myriad of potential routes, the problem would be overwhelming were it not for the ingenuity of the mathematicians and computer scientists who cracked it.

Sometimes when people say the shortest distance between two points is a straight line, they mean it figuratively, as a way of ridiculing nuance and affirming common sense. In other words, keep it simple. But battling obstacles can give rise to great beauty—so much so that in art, and in math, it's often more fruitful to impose constraints on ourselves. Think of haiku, or sonnets, or telling the story of your life in six words. The same is true of all the math that's been created to help you find the shortest way from here to there when you can't take the easy way out.

Two points. Many paths. Mathematical bliss.

MATH SWAGGERS WITH an intimidating air of certainty. Like a Mafia capo, it comes across as decisive, unyielding, and strong. It'll make you an argument you can't refuse.

But in private, math is occasionally insecure. It has doubts. It questions itself and isn't always sure it's right. Especially where infinity is concerned. Infinity can keep math up at night, worrying, fidgeting, feeling existential dread. For there have been times in the history of math when unleashing infinity wrought such mayhem, there were fears it might blow up the whole enterprise. And that would be bad for business.

In the HBO series *The Sopranos*, mob boss Tony Soprano consults a psychiatrist, seeking treatment for anxiety attacks, trying to understand why his mother wants to have him killed, that sort of thing. Beneath a tough exterior of certainty lies a very confused and frightened person.

In much the same way, calculus put itself on the couch just when it seemed to be at its most lethal. After decades of triumph, of mowing down all the problems that stood in its way, it started to become aware of something rotten at its core. The very things that had made it most successful — its brutal skill and fearlessness in manipulating infinite processes — were now threatening to destroy it. And the therapy that eventually helped it through this crisis came to be known, coincidentally, as analysis.

Here's an example of the kind of problem that worried the mathematicians of the 1700s. Consider the infinite series

$$1 - 1 + 1 - 1 + 1 - 1 + \cdots.$$

It's the numerical equivalent of vacillating forever, taking one step forward, one step back, one step forward, one step back, and so on, ad infinitum.

Does this series even make sense? And if so, what does it equal?

Disoriented by an infinitely long expression like this, an optimist might hope that some of the old rules—the rules forged from experience with *finite* sums—would still apply. For example, we know that $1 + 2 = 2 + 1$; when we add two or more numbers in a finite sum, we can always switch their order without changing the result: $a + b$ equals $b + a$ (the commutative law of addition). And when there are more than two terms, we can always insert parentheses with abandon, grouping the terms however we like, without affecting the ultimate answer. For instance, $(1 + 2) + 4 = 1 + (2 + 4)$; adding 1 and 2 first, then 4, gives the same answer as adding 2 and 4 first, then 1. This is called the associative law of addition. It works even if some numbers are being subtracted, as long as we remember that subtracting a number is the same as adding its negative. For example, consider a three-term version of the series above, and ask: What is $1 - 1 + 1$? We could view it as either $(1 - 1) + 1$ or $1 + (-1 + 1)$, where in that second set of parentheses we've added negative 1 instead of subtracting 1. Either way, the answer comes out to be 1.

But when we try to generalize these rules to *infinite* sums, a few unpleasant surprises lie in store for us. Look at the contradiction that occurs if we trot out the associative law and trustingly apply it to $1 - 1 + 1 - 1 + 1 - 1 + \cdots$. On the one hand, it appears we can annihilate the positive and negative 1s by pairing them off like so:

$$1 - 1 + 1 - 1 + 1 - 1 + \cdots$$
$$= (1 - 1) + (1 - 1) + (1 - 1) + \cdots$$
$$= 0 + 0 + 0 + \cdots$$
$$= 0.$$

On the other hand, we could just as well insert the parentheses like this and conclude that the sum is 1:

$$1 - 1 + 1 - 1 + 1 - 1 + \cdots = 1 + (-1 + 1) + (-1 + 1) + \cdots$$
$$= 1 + 0 + 0 + \cdots$$
$$= 1.$$

Neither argument seems more convincing than the other, so perhaps the sum is *both* 0 and 1? That proposition sounds absurd to us today, but at the time some mathematicians were comforted by its religious overtones. It reminded them of the theological assertion that God created the world from nothing. As the mathematician and priest Guido Grandi wrote in 1703, "By putting parentheses into the expression $1 - 1 + 1 - 1 + \cdots$ in different ways, I can, if I want, obtain 0 or 1. But then the idea of creation *ex nihilo* is perfectly plausible."

Nevertheless, it appears that Grandi favored a third value for the sum, different from either 0 or 1. Can you guess what he thought it should be? Think of what you'd say if you were kidding but trying to sound scholastic.

Right — Grandi believed the true sum was $\frac{1}{2}$. And far superior mathematicians, including Leibniz and Euler, agreed. There were several lines of reasoning that supported this compromise. The simplest was to notice that $1 - 1 + 1 - 1 + \cdots$ could be expressed in terms of itself, as follows. Let's use the letter S to denote the sum. Then by definition

$$S = 1 - 1 + 1 - 1 + \cdots.$$

Now leave the first 1 on the right-hand side alone and look

at all the other terms. They harbor their own copy of S, positioned to the right of that first 1 and subtracted from it:

$$S = 1 - 1 + 1 - 1 + \cdots$$
$$= 1 - (1 - 1 + 1 - \cdots)$$
$$= 1 - S.$$

So $S = 1 - S$ and therefore $S = \frac{1}{2}$.

The debate over the series $1 - 1 + 1 - 1 + \cdots$ raged for about 150 years, until a new breed of analysts put all of calculus and its infinite processes (limits, derivatives, integrals, infinite series) on a firm foundation, once and for all. They rebuilt the subject from the ground up, fashioning a logical structure as sound as Euclid's geometry.

Two of their key notions are partial sums and convergence. A partial sum is a running total. You simply add up a finite number of terms and then stop. For example, if we sum the first three terms of the series $1 - 1 + 1 - 1 + \cdots$, we get $1 - 1 + 1 = 1$. Let's call this S_3. Here the letter S stands for "sum" and the subscript 3 indicates that we added only the first three terms.

Similarly, the first few partial sums for this series are

$$S_1 = 1$$
$$S_2 = 1 - 1 = 0$$
$$S_3 = 1 - 1 + 1 = 1$$
$$S_4 = 1 - 1 + 1 - 1 = 0.$$

Thus we see that the partial sums bobble back and forth between 0 and 1, with no tendency to settle down to 0 or 1, to $\frac{1}{2}$, or to anything else. For this reason, mathematicians today would say that the series $1 - 1 + 1 - 1 + \cdots$ does not converge.

In other words, its partial sums don't approach *any* limiting value as more and more terms are included in the sum. Therefore the sum of the infinite series is meaningless.

So suppose we keep to the straight and narrow — no dallying with the dark side — and restrict our attention to only those series that converge. Does that get rid of the earlier paradoxes?

Not yet. The nightmares continue. And it's just as well that they do, because by facing down these new demons, the analysts of the 1800s discovered deeper secrets at the heart of calculus and then exposed them to the light. The lessons learned have proved invaluable, not just within math but for math's applications to everything from music to medical imaging.

Consider this series, known in the trade as the alternating harmonic series:

$$1 - \tfrac{1}{2} + \tfrac{1}{3} - \tfrac{1}{4} + \tfrac{1}{5} - \tfrac{1}{6} + \cdots .$$

Instead of one step forward, one step back, the steps now get progressively smaller. It's one step forward, but only *half* a step back, then a *third* of the step forward, a *fourth* of a step back, and so on. Notice the pattern: the fractions with odd denominators have plus signs in front of them, while the even fractions have negative signs. The partial sums in this case are

$$S_1 = 1$$
$$S_2 = 1 - \tfrac{1}{2} = 0.500$$
$$S_3 = 1 - \tfrac{1}{2} + \tfrac{1}{3} = 0.833 \ldots$$
$$S_4 = 1 - \tfrac{1}{2} + \tfrac{1}{3} - \tfrac{1}{4} = 0.583 \ldots$$

And if you go far enough, you'll find that they home in on a number close to 0.69. In fact, the series can be proven to converge. Its limiting value is the natural logarithm of 2, denoted ln2 and approximately equal to 0.693147.

So what's nightmarish here? On the face of it, nothing. The alternating harmonic series seems like a nice, well-behaved, convergent series, the sort your parents would approve of.

And that's what makes it so dangerous. It's a chameleon, a con man, a slippery sicko that will be anything you want. If you add up its terms in a different order, you can make it sum to anything. Literally. It can be rearranged to converge to *any* real number: 297.126, or −42π, or 0, or whatever your heart desires.

It's as if the series had utter contempt for the commutative law of addition. Merely by adding its terms in a different order, you can change the answer — something that could *never* happen for a finite sum. So even though the original series converges, it's still capable of weirdness unimaginable in ordinary arithmetic.

Rather than prove this astonishing fact (a result known as the Riemann rearrangement theorem), let's look at a particularly simple rearrangement whose sum is easy to calculate. Suppose we add *two* of the negative terms in the alternating harmonic series for every *one* of its positive terms, as follows:

$$[1 - \tfrac{1}{2} - \tfrac{1}{4}] + [\tfrac{1}{3} - \tfrac{1}{6} - \tfrac{1}{8}] + [\tfrac{1}{5} - \tfrac{1}{10} - \tfrac{1}{12}] + \cdots.$$

Next, simplify each of the bracketed expressions by subtracting the second term from the first while leaving the third term untouched. Then the series reduces to

$$[\tfrac{1}{2} - \tfrac{1}{4}] + [\tfrac{1}{6} - \tfrac{1}{8}] + [\tfrac{1}{10} - \tfrac{1}{12}] + \cdots.$$

After factoring out $\frac{1}{2}$ from all the fractions above and collecting terms, this becomes

$$\frac{1}{2}\left[1 - \frac{1}{2} + \frac{1}{3} - \frac{1}{4} + \frac{1}{5} - \frac{1}{6} + \cdots\right].$$

Look who's back: the beast inside the brackets is the alternating harmonic series itself. By rearranging it, we've somehow made it *half* as big as it was originally—even though it contains all the same terms! Arranged in this order, the series now converges to $\frac{1}{2}\ln 2 = 0.346\ldots$

Strange, yes. Sick, yes. And surprisingly enough, it matters in real life too. As we've seen throughout this book, even the most abstruse and far-fetched concepts of math often find application to practical things. The link in the present case is that in many parts of science and technology, from signal processing and acoustics to finance and medicine, it's useful to represent various kinds of curves, sounds, signals, or images as sums of simpler curves, sounds, signals, or images. When the basic building blocks are sine waves, the technique is known as Fourier analysis, and the corresponding sums are called Fourier series. But when the series in question bears some of the same pathologies as the alternating harmonic series and its equally deranged relatives, the convergence behavior of the Fourier series can be very weird indeed.

Here, for example, is a Fourier series directly inspired by the alternating harmonic series:

$$f(x) = \sin x - \frac{1}{2}\sin 2x + \frac{1}{3}\sin 3x - \frac{1}{4}\sin 4x + \cdots.$$

To get a sense for what this looks like, let's graph the sum of its first ten terms.

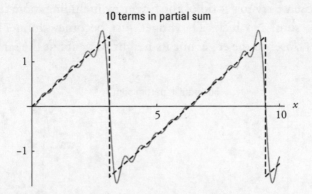

This partial sum (shown as a solid line) is clearly trying to approximate a much simpler curve, a wave shaped like the teeth of a saw (shown by the dashed line). Notice, however, that something goes wrong near the edges of the teeth. The sine waves overshoot the mark there and produce a strange finger that isn't in the sawtooth wave itself. To see this more clearly, here's a zoom near one of those edges, at $x = \pi$:

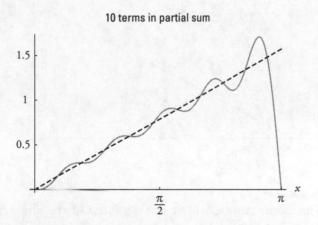

Suppose we try to get rid of the finger by including more terms in the sum. No luck. The finger just becomes thinner and moves closer to the edge, but its height stays about the same.

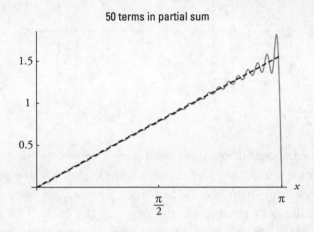

50 terms in partial sum

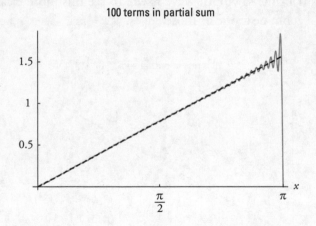

100 terms in partial sum

The blame can be laid at the doorstep of the alternating harmonic series. Its pathologies discussed earlier now contaminate the associated Fourier series. They're responsible for that annoying finger that just won't go away.

This effect, commonly called the Gibbs phenomenon, is more than a mathematical curiosity. Known since the mid-1800s, it now turns up in our digital photographs and on MRI scans. The unwanted oscillations caused by the Gibbs phenomenon can produce blurring, shimmering, and other artifacts at sharp edges in the image. In a medical context, these can be mistaken for damaged tissue, or they can obscure lesions that are actually present.

Fortunately, analysts a century ago pinpointed what causes Gibbs artifacts (see the notes on page 303 for discussion). Their insights have taught us how to overcome them, or at least how to spot them when they do occur.

The therapy has been very successful. The copay is due now.

In FEBRUARY 2010 I received an e-mail from a woman named Kim Forbes. Her six-year-old son, Ben, had asked her a math question she couldn't answer, and she was hoping I could help:

> Today is the 100th day of school. He was very excited and told me everything he knows about the number 100, including that 100 was an even number. He then told me that 101 was an odd number and 1 million was an even number, etc. He then paused and asked: "Is infinity even or odd?"

I explained to Kim that infinity is neither even nor odd. It's not a number in the usual sense, and it doesn't obey the rules of arithmetic. All sorts of contradictions would follow if it did. For instance, I wrote, "if infinity were odd, 2 times infinity would be even. But both are infinity! So the whole idea of odd and even docs not make sense for infinity."

Kim replied:

> Thank you. Ben was satisfied with that answer and
> kind of likes the idea that infinity is big enough to be
> both odd and even.

Although something got garbled in translation (infinity is *neither* odd nor even, not *both*), Ben's rendering hints at a larger truth. Infinity can be mind-boggling.

Some of its strangest aspects first came to light in the late 1800s, with Georg Cantor's groundbreaking work on set theory. Cantor was particularly interested in infinite sets of numbers and points, like the set $\{1, 2, 3, 4, \ldots\}$ of natural numbers and the set of points on a line. He defined a rigorous way to compare different infinite sets and discovered, shockingly, that some infinities are bigger than others.

At the time, Cantor's theory provoked not just resistance, but outrage. Henri Poincaré, one of the leading mathematicians of the day, called it a "disease." But another giant of the era, David Hilbert, saw it as a lasting contribution and later proclaimed, "No one shall expel us from the Paradise that Cantor has created."

My goal here is to give you a glimpse of this paradise. But rather than working directly with sets of numbers or points, let me follow an approach introduced by Hilbert himself. He vividly conveyed the strangeness and wonder of Cantor's theory by telling a parable about a grand hotel, now known as the Hilbert Hotel.

It's always booked solid, yet there's always a vacancy.

For the Hilbert Hotel doesn't have merely hundreds of rooms—it has an *infinite* number of them. Whenever a new guest arrives, the manager shifts the occupant of room 1 to room 2, room 2 to room 3, and so on. That frees up room 1 for the newcomer and accommodates everyone else as well (though inconveniencing them by the move).

Now suppose *infinitely* many new guests arrive, sweaty and short-tempered. No problem. The unflappable manager moves the occupant of room 1 to room 2, room 2 to room 4, room 3 to room 6, and so on. This doubling trick opens up all the odd-numbered rooms—infinitely many of them—for the new guests.

Later that night, an endless convoy of buses rumbles up to reception. There are infinitely many buses, and worse still, each one is loaded with an infinity of crabby people demanding that the hotel live up to its motto, "There's always room at the Hilbert Hotel."

The manager has faced this challenge before and takes it in stride.

First he does the doubling trick. That reassigns the current guests to the even-numbered rooms and clears out all the odd-numbered ones—a good start, because he now has an infinite number of rooms available.

But is that enough? Are there really enough odd-numbered rooms to accommodate the teeming horde of new guests? It seems unlikely, since there are something like infinity squared people clamoring for these rooms. (Why infinity squared? Because there was an infinite number of people on each of an infinite number of buses, and that amounts to infinity times infinity, whatever that means.)

This is where the logic of infinity gets very weird.

To understand how the manager is going to solve his latest problem, it helps to visualize all the people he has to serve.

Passengers

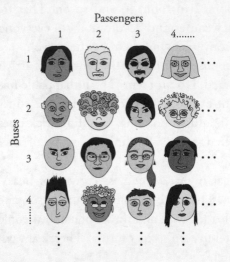

Of course, we can't show literally all of them here, since the diagram would need to be infinite in both directions. But a finite version of the picture is adequate. The point is that any *specific* bus passenger (your aunt Inez, say, on vacation from Louisville) is sure to appear on the diagram somewhere, as long as we include enough rows and columns. In that sense, everybody on every bus is accounted for. You name the passenger, and he or she is certain to be depicted at some finite number of steps east and south of the diagram's corner.

The manager's challenge is to find a way to work through this picture systematically. He needs to devise a scheme for assigning rooms so that everybody gets one eventually, after only a *finite* number of other people have been served.

Sadly, the previous manager hadn't understood this, and mayhem ensued. When a similar convoy showed up on his watch, he became so flustered trying to process all the people on bus 1 that he never got around to any other bus, leaving all those neglected passengers screaming and furious. Illustrated on the diagram below, this myopic strategy would correspond to a path that marched eastward along row 1 forever.

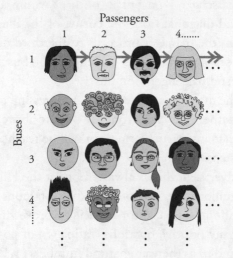

The new manager, however, has everything under control. Instead of tending to just one bus, he zigs and zags through the diagram, fanning out from the corner, as shown below.

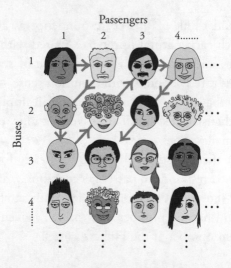

He starts with passenger 1 on bus 1 and gives her the first empty room. The second and third empty rooms go to passenger 2 on bus 1 and passenger 1 on bus 2, both of whom are depicted on the second diagonal from the corner of the diagram. After serving them, the manager proceeds to the third diagonal and hands out a set of room keys to passenger 1 on bus 3, passenger 2 on bus 2, and passenger 3 on bus 1.

I hope the manager's procedure — progressing from one diagonal to another — is clear from the picture above, and you're convinced that any particular person will be reached in a finite number of steps.

So, as advertised, there's always room at the Hilbert Hotel.

The argument I've just presented is a famous one in the theory of infinite sets. Cantor used it to prove that there are exactly as many positive fractions (ratios p/q of positive whole numbers p and q) as there are natural numbers $(1, 2, 3, 4, \ldots)$. That's a much stronger statement than saying both sets are infinite. It says they are infinite to precisely the same extent, in the sense that a one-to-one correspondence can be established between them.

You could think of this correspondence as a buddy system in which each natural number is paired with some positive fraction, and vice versa. The existence of such a buddy system seems utterly at odds with common sense — it's the sort of sophistry that made Poincaré recoil. For it implies we could make an exhaustive list of all positive fractions, even though there's no smallest one!

And yet there is such a list. We've already found it. The fraction p/q corresponds to passenger p on bus q, and the argument above shows that each of these fractions can be paired off with a certain natural number $1, 2, 3, \ldots$, given by the passenger's room number at the Hilbert Hotel.

The coup de grâce is Cantor's proof that some infinite sets are bigger than this. Specifically, the set of real numbers between 0 and 1 is uncountable — it can't be put in one-to-one correspondence with the natural numbers. For the hospitality industry, this means that if all these real numbers show up at the reception desk and bang on the bell, there won't be enough rooms for all of them, even at the Hilbert Hotel.

The proof is by contradiction. Suppose each real number could be given its own room. Then the roster of occupants, identified by their decimal expansions and listed by room number, would look something like this:

        Room 1:   .6708112345 . . .
        Room 2:   .1918676053 . . .
        Room 3:   .4372854675 . . .
        Room 4:   .2845635480 . . .

Remember, this is supposed to be a complete list. Every real number between 0 and 1 is supposed to appear somewhere, at some finite place on the roster.

Cantor showed that a lot of numbers are missing from any such list; that's the contradiction. For instance, to construct one that appears nowhere on the list shown above, go down the diagonal and build a new number from the underlined digits:

        Room 1:   .<u>6</u>708112345 . . .
        Room 2:   .1<u>9</u>18676053 . . .
        Room 3:   .43<u>7</u>2854675 . . .
        Room 4:   .284<u>5</u>635480 . . .

The decimal so generated is .6975 . . .

But we're not done yet. The next step is to take this decimal and change all its digits, replacing each of them with any *other* digit between 1 and 8. For example, we could change the 6 to a 3, the 9 to a 2, the 7 to a 5, and so on.

This new decimal .325 . . . is the killer. It's certainly not in room 1, since it has a different first digit from the number there. It's also not in room 2, since its second digit disagrees. In general, it differs from the nth number in the nth decimal place. So it doesn't appear anywhere on the list!

The conclusion is that the Hilbert Hotel can't accommodate all the real numbers. There are simply too many of them, an infinity beyond infinity.

And with that humbling thought, we come to the end of this book, which began with a scene in another imaginary hotel. A *Sesame Street* character named Humphrey, working the lunch shift at the Furry Arms, took an order from a roomful of hungry penguins—"Fish, fish, fish, fish, fish, fish"—and soon learned about the power of numbers.

It's been a long journey from fish to infinity. Thanks for joining me.

Acknowledgments

Many friends and colleagues helped improve this book by generously offering their sage advice — mathematical, stylistic, historical, and otherwise. Thanks to Doug Arnold, Sheldon Axler, Larry Braden, Dan Callahan, Bob Connelly, Tom Gilovich, George Hart, Vi Hart, Diane Hopkins, Herbert Hui, Cindy Klauss, Michael Lewis, Michael Mauboussin, Barry Mazur, Eri Noguchi, Charlie Peskin, Steve Pinker, Ravi Ramakrishna, David Rand, Richard Rand, Peter Renz, Douglas Rogers, John Smillie, Grant Wiggins, Stephen Yeung, and Carl Zimmer.

Other colleagues created images for this book or allowed me to include their visual work. Thanks to Rick Allmendinger, Paul Bourke, Mike Field, Brian Madsen, Nik Dayman (Teamfresh), Mark Newman, Konrad Polthier, Christian Rudder at OkCupid, Simon Tatham, and Jane Wang.

I am immensely grateful to David Shipley for inviting me to write the *New York Times* series that led to this book, and especially for his vision of how the series should be structured. Simplicity, simplicity, simplicity, urged Thoreau — and both he and Shipley were right. George Kalogerakis, my editor at the *Times*, wielded his pen lightly, moving commas, but only when necessary, while protecting me from more serious infelicities.

His confidence was enormously reassuring. Katie O'Brien on the production team made sure the math always looked right and put up with the requisite typographical fussing with grace and good humor.

I feel so fortunate to have Katinka Matson in my corner as my literary agent. She championed this book from the beginning with an enthusiasm that was inspirational.

Paul Ginsparg, Jon Kleinberg, Tim Novikoff, and Andy Ruina read drafts of nearly every chapter, their only compensation being the pleasure of catching clunkers and using their brilliant minds for good instead of evil. Normally it's a drag to be around such know-it-alls, but the fact is, they *do* know it all. And this book is the better for it. I'm truly grateful for their effort and encouragement.

Thanks to Margy Nelson, the illustrator, for her playfulness and scientific sensibility. She often felt to me like a partner in this project, with her knack for finding original ways to convey the essence of a mathematical concept.

Any writer would be blessed to have Amanda Cook as an editor. How can anyone be so gentle and wise and decisive, all at the same time? Thank you, Amanda, for believing in this book and for helping me shape every bit of it. Eamon Dolan, another of the world's great editors, guided this project (and me) toward the finish line with a sure hand and infectious excitement. Editorial assistants Ashley Gilliam and Ben Hyman were meticulous and fun to work with and took good care of the book at every stage of its development. Copyeditor Tracy Roe taught me about appositives, apostrophes, and words as words. But more important (not "importantly"!), she sharpened the writing and thinking in these pages. Thanks as well to publicist Michelle Bonanno, marketing manager Ayesha Mirza, production editor Rebecca Springer, production manager David Futato, and the entire team at Houghton Mifflin Harcourt.

Finally, let me add my most heartfelt thanks to my family. Leah and Jo, you've been hearing about the book for a long time now, and believe it or not, it really has come to an end. Your next chore, naturally, is to learn all the math in it. And as for my fantastically patient wife, Carole, who slogged through the first n drafts of each chapter and thereby learned the true meaning of the expression "as n tends to infinity," let me say simply, I love you. Finding you was the best problem I ever solved.

Notes

1. From Fish to Infinity

3   Sesame Street: The video *Sesame Street: 123 Count with Me* (1997) is available for purchase online in either VHS or DVD format.

4   *numbers . . . have lives of their own*: For a passionate presentation of the idea that numbers have lives of their own and the notion that mathematics can be viewed as a form of art, see P. Lockhart, *A Mathematician's Lament* (Bellevue Literary Press, 2009).

5   *"the unreasonable effectiveness of mathematics"*: The essay that introduced this now-famous phrase is E. Wigner, "The unreasonable effectiveness of mathematics in the natural sciences," *Communications in Pure and Applied Mathematics*, Vol. 13, No. 1 (February 1960), pp. 1–14. An online version is available at http://www.dartmouth.edu/~matc/MathDrama/reading/Wigner.html.

For further reflections on these ideas and on the related question of whether math was invented or discovered, see M. Livio, *Is God a Mathematician?* (Simon and Schuster, 2009), and R. W. Hamming, "The unreasonable effectiveness of mathematics," *American Mathematical Monthly*, Vol. 87, No. 2 (February 1980), available online at http://www.-lmmb.ncifcrf.gov/~toms/Hamming.unreasonable.html.

2. Rock Groups

7   *The playful side of arithmetic*: As I hope to make clear, this chapter owes much to two books—one a polemic, the other a novel, both

of them brilliant: P. Lockhart, *A Mathematician's Lament* (Bellevue Literary Press, 2009), which inspired the rock metaphor and some of the examples used here; and Y. Ogawa, *The Housekeeper and the Professor* (Picador, 2009).

a child's curiosity: For young readers who like exploring numbers and the patterns they make, see H. M. Enzensberger, *The Number Devil* (Holt Paperbacks, 2000).

10 *hallmark of an elegant proof*: Delightful but more advanced examples of visualization in mathematics are presented in R. B. Nelsen, *Proofs without Words* (Mathematical Association of America, 1997).

3. The Enemy of My Enemy

17 *"Yeah, yeah"*: For more of Sidney Morgenbesser's witticisms and academic one-liners, see the sampling at Language Log (August 5, 2004), "If P, so why not Q?" online at http://itre.cis.upenn .edu/%7Emyl/languagelog/archives/001314.html.

18 *relationship triangles*: Balance theory was first proposed by the social psychologist Fritz Heider and has since been developed and applied by social network theorists, political scientists, anthropologists, mathematicians, and physicists. For the original formulation, see F. Heider, "Attitudes and cognitive organization," *Journal of Psychology*, Vol. 21 (1946), pp. 107–112, and F. Heider, *The Psychology of Interpersonal Relations* (John Wiley and Sons, 1958). For a review of balance theory from a social network perspective, see S. Wasserman and K. Faust, *Social Network Analysis* (Cambridge University Press, 1994), chapter 6.

19 *polarized states are the* only *states as stable as nirvana*: The theorem that a balanced state in a fully connected network must be either a single nirvana of all friends or two mutually antagonistic factions was first proven in D. Cartwright and F. Harary, "Structural balance: A generalization of Heider's theory," *Psychological Review*, Vol. 63 (1956), pp. 277–293. A very readable version of that proof, and a gentle introduction to the mathematics of balance theory, has

been given by two of my colleagues at Cornell: D. Easley and J. Kleinberg, *Networks, Crowds, and Markets* (Cambridge University Press, 2010).

In much of the early work on balance theory, a triangle of three mutual enemies (and hence three negative sides) was considered unbalanced. I assumed this implicitly when quoting the results about nirvana and the two-bloc state being the only configurations of a fully connected network in which all triangles are balanced. However, some researchers have challenged this assumption and have explored the implications of treating a triangle of three negatives as balanced. For more on this and other generalizations of balance theory, see the books by Wasserman and Faust and by Easley and Kleinberg cited above.

World War I: The example and graphical depiction of the shifting alliances before World War I are from T. Antal, P. L. Krapivsky, and S. Redner, "Social balance on networks: The dynamics of friendship and enmity," *Physica D*, Vol. 224 (2006), pp. 130–136, available online at http://arxiv.org/abs/physics/0605183. This paper, written by three statistical physicists, is notable for recasting balance theory in a dynamic framework, thus extending it beyond the earlier static approaches. For the historical details of the European alliances, see W. L. Langer, *European Alliances and Alignments, 1871–1890*, 2nd edition (Knopf, 1956), and B. E. Schmitt, *Triple Alliance and Triple Entente* (Henry Holt and Company, 1934).

4. Commuting

23 *revisit multiplication from scratch*: Keith Devlin has written a provocative series of essays about the nature of multiplication: what it is, what it is not, and why certain ways of thinking about it are more valuable and reliable than others. He argues in favor of thinking of multiplication as scaling, not repeated addition, and shows that the two concepts are very different in real-world settings where units are involved. See his January 2011 blog post "What exactly is multiplication?" at http://www.maa.org/devlin/devlin_01_11.html, as well as three earlier posts from 2008: "It ain't no repeated addition" (http://www.maa.org/devlin/devlin_06_08.html); "It's still not

repeated addition" (http://www.maa.org/devlin/devlin_0708_08
.html); and "Multiplication and those pesky British spellings"
(http://www.maa.org/devlin/devlin_09_08.html). These essays
generated a lot of discussion in the blogosphere, especially among
schoolteachers. If you're short on time, I'd recommend reading the
one from 2011 first.

25 *shopping for a new pair of jeans*: For the jeans example, the order in
which the tax and discount are applied may not matter to you—in
both scenarios you end up paying $43.20—but it makes a big
difference to the government and the store! In the clerk's scenario
(where you pay tax based on the original price), you would pay
$4 in tax; in your scenario, only $3.20. So how can the final price
come out the same? It's because in the clerk's scenario the store gets
to keep $39.20, whereas in yours it would keep $40. I'm not sure
what the law requires, and it may vary from place to place, but the
rational thing would be for the government to charge sales tax based
on the actual payment the store receives. Only your scenario satisfies this criterion. For further discussion, see http://www.facebook
.com/TeachersofMathematics/posts/166897663338316.

26 *financial decisions*: For heated online arguments about the relative merits of a Roth 401(k) versus a traditional one, and
whether the commutative law has anything to do with these issues, see the Finance Buff, "Commutative law of multiplication"
(http://thefinancebuff.com/commutative-law-of-multiplication
.html), and the Simple Dollar, "The new Roth 401(k) versus the
traditional 401(k): Which is the better route?" (http://www
.thesimpledollar.com/2007/06/20/the-new-roth-401k-versus-the-
traditional-401k-which-is-the-better-route/).

27 *attending MIT and killing himself didn't commute*: This story about
Murray Gell-Mann is recounted in G. Johnson, *Strange Beauty*
(Knopf, 1999), p. 55. In Gell-Mann's own words, he was offered
admission to the "dreaded" Massachusetts Institute of Technology
at the same time as he was "contemplating suicide, as befits someone rejected from the Ivy League. It occurred to me however (and it

is an interesting example of non-commutation of operators) that I could try M.I.T. first and kill myself later, while the reverse order of events was impossible." This excerpt appears in H. Fritzsch, *Murray Gell-Mann: Selected Papers* (World Scientific, 2009), p. 298.

development of quantum mechanics: For an account of how Heisenberg and Dirac discovered the role of non-commuting variables in quantum mechanics, see G. Farmelo, *The Strangest Man* (Basic Books, 2009), pp. 85–87.

5. Division and Its Discontents

30 My Left Foot: A clip of the scene where young Christy struggles valiantly to answer the question "What's twenty-five percent of a quarter?" is available online at http://www.tcm.com/mediaroom/ video/223343/My-Left-Foot-Movie-Clip-25-Percent-of-a-Quarter .html.

32 *Verizon Wireless*: George Vaccaro's blog (http://verizonmath .blogspot.com/) provides the exasperating details of his encounters with Verizon. The transcript of his conversation with customer service is available at http://verizonmath.blogspot.com/2006/12/ transcription-jt.html. The audio recording is at http://imgs.xkcd .com/verizon_billing.mp3.

33 *you're forced to conclude that 1 must equal .9999 . . .* : For readers who may still find it hard to accept that 1 = .9999 . . . , the argument that eventually convinced me was this: they must be equal, because there's no room for any other decimal to fit between them. (Whereas if two decimals are unequal, their average is between them, as are infinitely many other decimals.)

34 *almost all decimals are irrational*: The amazing properties of irrational numbers are discussed at a higher mathematical level on the MathWorld page "Irrational Number," http://mathworld .wolfram.com/IrrationalNumber.html. The sense in which the digits of irrational numbers are random is clarified at http://mathworld .wolfram.com/NormalNumber.html.

6. Location, Location, Location

35 *Ezra Cornell's statue*: For more about Cornell, including his role
in Western Union and the early days of the telegraph, see P. Dorf,
The Builder: A Biography of Ezra Cornell (Macmillan, 1952); W. P.
Marshall, *Ezra Cornell* (Kessinger Publishing, 2006); and http://
rmc.library.cornell.edu/ezra/index.html, an online exhibition in
honor of Cornell's 200th birthday.

36 *systems for writing numbers*: Ancient number systems and the ori-
gins of the decimal place-value system are discussed in V. J. Katz,
A History of Mathematics, 2nd edition (Addison Wesley Longman,
1998), and in C. B. Boyer and U. C. Merzbach, *A History of
Mathematics*, 3rd edition (Wiley, 2011). For a chattier account, see
C. Seife, *Zero* (Viking, 2000), chapter 1.

37 *Roman numerals*: Mark Chu-Carroll clarifies some of the peculiar features of Roman numerals and arithmetic in this blog post:
http://scienceblogs.com/goodmath/2006/08/roman_numerals_
and_arithmetic.php.

Babylonians: A fascinating exhibition of Babylonian math is described by N. Wade, "An exhibition that gets to the (square) root
of Sumerian math," *New York Times* (November 22, 2010), online at http://www.nytimes.com/2010/11/23/science/23babylon
.html, accompanied by a slide show at http://www.nytimes.com/
slideshow/2010/11/18/science/20101123-babylon.html.

nothing to do with human appendages: This may well be an overstate-
ment. You can count to twelve on one hand by using your thumb
to indicate each of the three little finger bones (phalanges) on the
other four fingers. Then you can use all five fingers on your other
hand to keep track of how many sets of twelve you've counted. The
base 60 system used by the Sumerians may have originated in this
way. For more on this hypothesis and other speculations about the
origins of the base 60 system, see G. Ifrah, *The Universal History of
Numbers* (Wiley, 2000), chapter 9.

7. The Joy of *x*

46 *Jo realized something about her big sister, Leah*: For sticklers, Leah is actually twenty-one months older than Jo. Hence Jo's formula is only an approximation. Obviously!

47 *"When I was at Los Alamos"*: Feynman tells the story of Bethe's trick for squaring numbers close to 50 in R. P. Feynman, *"Surely You're Joking, Mr. Feynman!"* (W. W. Norton and Company, 1985), p. 193.

49 *money invested in the stock market*: The identity about the effect of equal up-and-down percentage swings in the stock market can be proven symbolically, by multiplying $1 + x$ by $1 - x$, or geometrically, by drawing a diagram similar to that used to explain Bethe's trick. If you're in the mood, try both approaches as an exercise.

50 *socially acceptable age difference in a romance*: The "half your age plus seven" rule about the acceptable age gap in a romantic relationship is called the standard creepiness rule in this xkcd comic: http://xkcd.com/314/.

8. Finding Your Roots

51 *their struggle to find the roots*: The quest for solutions to increasingly complicated equations, from quadratic to quintic, is recounted in vivid detail in M. Livio, *The Equation That Couldn't Be Solved* (Simon and Schuster, 2005).

doubling a cube's volume: For more about the classic problem of doubling the cube, see http://www-history.mcs.st-and.ac.uk/HistTopics/Doubling_the_cube.html.

square roots of negative numbers: To learn more about imaginary and complex numbers, their applications, and their checkered history, see P. J. Nahin, *An Imaginary Tale* (Princeton University Press, 1998), and B. Mazur, *Imagining Numbers* (Farrar, Straus and Giroux, 2003).

55 *dynamics of Newton's method*: For a superb journalistic account of John Hubbard's work, see J. Gleick, *Chaos* (Viking, 1987), p. 217. Hubbard's own take on Newton's method appears in section 2.8 of J. Hubbard and B. B. Hubbard, *Vector Calculus, Linear Algebra, and Differential Forms*, 4th edition (Matrix Editions, 2009).

For readers who want to delve into the mathematics of Newton's method, a more sophisticated but still readable introduction is given in H.-O. Peitgen and P. H. Richter, *The Beauty of Fractals* (Springer, 1986), chapter 6, and also see the article by A. Douady (Hubbard's collaborator) entitled "Julia sets and the Mandelbrot set," starting on p. 161 of the same book.

57 *The borderlands looked like psychedelic hallucinations*: Hubbard was not the first mathematician to ask questions about Newton's method in the complex plane; Arthur Cayley had wondered about the same things in 1879. He too looked at both quadratic and cubic polynomials and realized that the first case was easy and the second was hard. Although he couldn't have known about the fractals that would be discovered a century later, he clearly understood that something nasty could happen when there were more than two roots. In his one-page article "Desiderata and suggestions: No. 3—the Newton-Fourier imaginary problem," *American Journal of Mathematics*, Vol. 2, No. 1 (March 1879), p. 97, available online at http://www.jstor.org/pss/2369201, Cayley's final sentence is a marvel of understatement: "The solution is easy and elegant in the case of a quadric equation, but the next succeeding case of the cubic equation appears to present considerable difficulty."

The structure was a fractal: The snapshots shown in this chapter were computed using Newton's method applied to the polynomial $z^3 - 1$. The roots are the three cube roots of 1. For this case, Newton's algorithm takes a point z in the complex plane and maps it to a new point

$$z - (z^3 - 1)/(3z^2).$$

That point then becomes the next z. This process is repeated until z comes sufficiently close to a root or, equivalently, until $z^3 - 1$ comes sufficiently close to zero, where "sufficiently close" is a very small distance, arbitrarily chosen by the person who programmed the computer. All initial points that lead to a particular root are then assigned the same color. Thus red labels all the points that converge to one root, green labels another, and blue labels the third.

The snapshots of the resulting Newton fractal were kindly provided by Simon Tatham. For more on his work, see his webpage "Fractals derived from Newton-Raphson iteration" (http://www .chiark.greenend.org.uk/~sgtatham/newton/).

Video animations of the Newton fractal have been created by Teamfresh. Stunningly deep zooms into other fractals, including the famous Mandelbrot set, are available at the Teamfresh website, http://www.hd-fractals.com.

58 *temple builders in India*: For an introduction to the ancient Indian methods for finding square roots, see D. W. Henderson and D. Taimina, *Experiencing Geometry*, 3rd expanded and revised edition (Pearson Prentice Hall, 2005).

9. My Tub Runneth Over

59 *my first word problem*: A large collection of classic word problems is available at http://MathNEXUS.wwu.edu/Archive/oldie/list.asp.

filling a bathtub: A more difficult bathtub problem appears in the 1941 movie *How Green Was My Valley*. For a clip, see http:// www.math.harvard.edu/~knill/mathmovies/index.html. And while you're there, you should also check out this clip from the baseball comedy *Little Big League:* http://www.math.harvard.edu/~knill/ mathmovies/m4v/league.m4v. It contains a word problem about painting houses: "If I can paint a house in three hours, and you can paint it in five, how long will it take us to paint it together?" The scene shows the baseball players giving various silly answers. "It's simple, five times three, so that's fifteen." "No, no, no, look it. It takes eight hours: five plus three, that's eight." After a few more blunders, one player finally gets it right: $1\frac{7}{8}$ hours.

10. Working Your Quads

67 *top ten most beautiful or important equations*: For books about great

equations, see M. Guillen, *Five Equations That Changed the World* (Hyperion, 1995); G. Farmelo, *It Must Be Beautiful* (Granta, 2002); and R. P. Crease, *The Great Equations* (W. W. Norton and Company, 2009). There are several lists posted online too. I'd suggest starting with K. Chang, "What makes an equation beautiful?," *New York Times* (October 24, 2004), http://www.nytimes.com/2004/10/24/weekinreview/24chan.html. For one of the few lists that include the quadratic equation, see http://www4.ncsu.edu/~kaltofen/top10eqs/top10eqs.html.

68 *calculating inheritances*: Many examples are discussed in S. Gandz, "The algebra of inheritance: A rehabilitation of al-Khuwarizmi," *Osiris*, Vol. 5 (1938), pp. 319–391.

69 *al-Khwarizmi*: Al-Khwarizmi's approach to the quadratic equation is explained in V. J. Katz, *A History of Mathematics*, 2nd edition (Addison Wesley Longman, 1998), pp. 244–249.

11. Power Tools

75 Moonlighting: The banter about logarithms is from the episode "In God We Strongly Suspect." It originally aired on February 11, 1986, during the show's second season. A video clip is available at http://opinionator.blogs.nytimes.com/2010/03/28/power-tools/.

76 *functions*: For simplicity, I've referred to expressions like x^2 as functions, though to be more precise I should speak of "the function that maps x into x^2." I hope this sort of abbreviation won't cause confusion, since we've all seen it on calculator buttons.

78 *breathtaking parabolas*: For a promotional video of the water feature at the Detroit airport, created by WET Design, see http://www.youtube.com/watch?v=VSUKNxVXE4E. Several home videos of it are also available at YouTube. One of the most vivid is "Detroit Airport Water Feature" by PassTravelFool (http://www.youtube.com/watch?v=or8i_EvIRdE).

Will Hoffman and Derek Paul Boyle have filmed an intriguing video of the parabolas (along with their exponential cousins, curves called catenaries, so named for the shape of hanging chains) that are all around us in the everyday world. See "WNYC/NPR's Radio Lab presents Parabolas (etc.)" online at http://www.youtube .com/watch?v=rdSgqHuI-mw. Full disclosure: the filmmakers say this video was inspired by a story I told on an episode of *Radiolab* ("Yellow fluff and other curious encounters," available at http:// www.radiolab.org/2009/jan/12/).

79 *fold a piece of paper in half more than seven or eight times*: For the story of Britney Gallivan's adventures in paper folding, see B. Gallivan, "How to fold a paper in half twelve times: An 'impossible challenge' solved and explained," Pomona, CA: Historical Society of Pomona Valley, 2002, online at http://pomonahistorical .org/12times.htm. For a journalist's account aimed at children, see I. Peterson, "Champion paper-folder," *Muse* (July/August 2004), p. 33, available online at http://musemath.blogspot.com/2007/06/ champion-paper-folder.html. The MythBusters have attempted to replicate Britney's experiment on their television show (http://kwc .org/mythbusters/2007/01/episode_72_underwater_car_and.html).

81 *We perceive pitch logarithmically*: For references and further discussion of musical scales and our (approximately) logarithmic perception of pitch, see J. H. McDermott and A. J. Oxenham, "Music perception, pitch, and the auditory system," *Current Opinion in Neurobiology*, Vol. 18 (2008), pp. 1–12; http://en.wikipedia.org/ wiki/Pitch_(music); http://en.wikipedia.org/wiki/Musical_scale; and http://en.wikipedia.org/wiki/Piano_key_frequencies.

For evidence that our innate number sense is also logarithmic, see S. Dehaene, V. Izard, E. Spelke, and P. Pica, "Log or linear? Distinct intuitions of the number scale in Western and Amazonian indigene cultures," *Science*, Vol. 320 (2008), pp. 1217–1220. Popular accounts of this study are available at ScienceDaily (http:// www.sciencedaily.com/releases/2008/05/080529141344.htm) and in an episode of *Radiolab* called "Numbers" (http://www.radiolab .org/2009/nov/30/).

12. Square Dancing

85 *Pythagorean theorem*: The ancient Babylonians, Indians, and Chinese appear to have been aware of the content of the Pythagorean theorem several centuries before Pythagoras and the Greeks. For more about the history and significance of the theorem, as well as a survey of the many ingenious ways to prove it, see E. Maor, *The Pythagorean Theorem* (Princeton University Press, 2007).

86 *hypotenuse*: On p. xiii of his book, Maor explains that the word "hypotenuse" means "stretched beneath" and points out that this makes sense if the right triangle is viewed with its hypotenuse at the bottom, as depicted in Euclid's *Elements*. He also notes that this interpretation fits well with the Chinese word for hypotenuse: "*hsien*, a string stretched between two points (as in a lute)."

88 *building the squares out of many little crackers*: Children and their parents will enjoy the edible illustrations of the Pythagorean theorem suggested by George Hart in his post for the Museum of Mathematics "Pythagorean crackers," http://momath.org/home/pythagorean-crackers/.

90 *another proof*: If you enjoy seeing different proofs, a nicely annotated collection of dozens of them—with creators ranging from Euclid to Leonardo da Vinci to President James Garfield—is available at Alexander Bogomolny's blog Cut the Knot. See http://www.cut-the-knot.org/pythagoras/index.shtml.

92 *missing steps*: With any luck, the first proof in the chapter should have given you an Aha! sensation. But to make the argument completely airtight, we also need to prove that the pictures aren't deceiving us—in other words, they truly have the properties they appear to have. A more rigorous proof would establish, for example, that the outer frame is truly a square, and that the medium and small squares meet at a single point, as shown. Checking these details is fun and not too difficult.

Here are the missing steps in the second proof. Take this equation:

$$a/d = c/a$$

and rearrange it to get

$$d = a^2/c.$$

Similarly massaging another of the equations yields

$$e = b^2/c.$$

Finally, substituting the expressions above for d and e into the equation $c = d + e$ yields

$$c = a^2/c + b^2/c.$$

Then multiplying both sides by c gives the desired formula:

$$c^2 = a^2 + b^2.$$

13. Something from Nothing

93 *Euclid*: For all thirteen books of the *Elements* in a convenient one-volume format, with plentiful diagrams, see *Euclid's Elements*, edited by D. Densmore, translated by T. L. Heath (Green Lion Press, 2002). Another excellent option is a free downloadable PDF document by Richard Fitzpatrick giving his own modern translation of Euclid's *Elements*, available at http://farside.ph.utexas.edu/euclid .html.

94 *Thomas Jefferson*: For more about Thomas Jefferson's reverence for Euclid and Newton and his use of their axiomatic approach in the Declaration of Independence, see I. B. Cohen, *Science and the Founding Fathers* (W. W. Norton and Company, 1995), pp. 108–134, as well as J. Fauvel, "Jefferson and mathematics," http:// www.math.virginia.edu/Jefferson/jefferson.htm, especially the page about the Declaration of Independence: http://www.math.virginia .edu/Jefferson/jeff_r(4).htm.

95 *equilateral triangle*: For Euclid's version of the equilateral triangle proof, in Greek, see http://en.wikipedia.org/wiki/File:Euclid-proof .jpg.

100 *logical* and *creative*: I have glossed over a number of subtleties in the two proofs presented in this chapter. For example, in the equilateral

triangle proof, we implicitly assumed (as Euclid did) that the two circles intersect somewhere—in particular, at the point we labeled C. But the existence of that intersection is not guaranteed by any of Euclid's axioms; one needs an additional axiom about the continuity of the circles. Bertrand Russell, among others, noted this lacuna: B. Russell, "The Teaching of Euclid," *Mathematical Gazette*, Vol. 2, No. 33 (1902), pp. 165–167, available online at http://www-history.mcs.st-and.ac.uk/history/Extras/Russell_Euclid.html.

Another subtlety involves the implicit use of the parallel postulate in the proof that the angles of a triangle sum to 180 degrees. That postulate is what gave us permission to construct the line parallel to the triangle's base. In other types of geometry (known as non-Euclidean geometries), there might exist *no* line parallel to the base, or *infinitely many* such lines. In those kinds of geometries, which are every bit as logically consistent as Euclid's, the angles of a triangle *don't* always add up to 180 degrees. Thus, the Pythagorean proof given here is more than just stunningly elegant; it reveals something deep about the nature of space itself. For more commentary on these issues, see the blog post by A. Bogomolny, "Angles in triangle add to 180°," http://www.cut-the-knot.org/triangle/pythpar/AnglesInTriangle.shtml, and the article by T. Beardon, "When the angles of a triangle don't add up to 180 degrees," http://nrich.maths.org/1434.

14. The Conic Conspiracy

104 *parabolas and ellipses*: For background on conic sections and references to the vast literature on them, see http://mathworld.wolfram.com/ConicSection.html and http://en.wikipedia.org/wiki/Conic_section. For readers with some mathematical training, a large amount of interesting and unusual information has been collected by James B. Calvert on his website; see "Ellipse" (http://mysite.du.edu/~jcalvert/math/ellipse.htm), "Parabola" (http://mysite.du.edu/~jcalvert/math/parabola.htm), and "Hyperbola" (http://mysite.du.edu/~jcalvert/math/hyperb.htm).

106 *parabolic mirror*: You'll gain a lot of intuition by watching the online animations created by Lou Talman and discussed on his page "The

geometry of the conic sections," http://rowdy.mscd.edu/~talmanl/ HTML/GeometryOfConicSections.html. In particular, watch http://clem.mscd.edu/~talmanl/HTML/ParabolicReflector.html and fix on a single photon as it approaches and then bounces off the parabolic reflector. Then look at all the photons moving together. You'll never want to tan your face with a sun reflector again. The analogous animation for an ellipse is shown at http://rowdy.mscd .edu/~talmanl/HTML/EllipticReflector.html.

15. Sine Qua Non

113 *sunrises and sunsets*: The charts shown in the text are for Jupiter, Florida, using data from 2011. For convenience, the times of sunrise and sunset have been expressed relative to Eastern Standard Time (the UTC-05:00 time zone) all year long to avoid the artificial discontinuities caused by daylight-saving time. You can create similar charts of sunrise and sunset for your own location at websites such as http://ptaff.ca/soleil/?lang=en_CA or http://www.gaisma .com/en/.

Students seem to find these charts surprising (for instance, some of them expect the curves to be triangular in appearance instead of rounded and smooth), which can make for instructive classroom activities at the high-school or middle-school level. For a pedagogical case study, see A. Friedlander and T. Resnick, "Sunrise, sunset," *Montana Mathematics Enthusiast*, Vol. 3, No. 2 (2006), pp. 249–255, available at http://www.math.umt.edu/tmme/vol3no2/ TMMEvol3no2_Israel_pp249_255.pdf.

Deriving formulas for the times of sunrise and sunset is complicated, both mathematically and in terms of the physics involved. See, for example, T. L. Watts's webpage "Variation in the time of sunrise" at http://www.physics.rutgers.edu/~twatts/sunrise/sunrise .html. Watts's analysis clarifies why the times of sunrise and sunset do not vary as simple sine waves throughout the year. They also include a second harmonic (a sine wave with a period of six months), mainly due to a subtle effect of the Earth's tilt that causes a semiannual variation in local noon, the time of day when the sun is highest in the sky. Happily, this term is the same in the formulas for both sunrise and sunset times. So when you subtract one from the other to compute the length of the day (the number of hours between sunrise and sunset), the second harmonic cancels out. What's left is very nearly a perfect sine wave.

More information about all this can be found on the Web by searching for "the Equation of Time." (Seriously—that's what it's called!) A good starting point is K. Taylor's webpage "The equation of time: Why sundial time differs from clock time depending on time of year," http://myweb.tiscali.co.uk/moonkmft/Articles/EquationOfTime.html, or the Wikipedia page http://en.wikipedia.org/wiki/Equation_of_time.

114 *trigonometry*: The subject is lovingly surveyed in E. Maor, *Trigonometric Delights* (Princeton University Press, 1998).

117 *pattern formation*: For a broad overview of patterns in nature, see P. Ball, *The Self-Made Tapestry* (Oxford University Press, 1999). The mathematical methods in this field are presented at a graduate level in R. Hoyle, *Pattern Formation* (Cambridge University Press, 2006). For mathematical analyses of zebra stripes, butterfly-wing patterns, and other biological examples of pattern formation, see J. D. Murray, *Mathematical Biology: II. Spatial Models and Biomedical Applications*, 3rd edition (Springer, 2003).

119 *cosmic microwave background*: The connections between biological pattern formation and cosmology are one of the many delights to be found in Janna Levin's book *How the Universe Got Its Spots* (Princeton University Press, 2002). It's structured as a series of unsent letters to her mother and ranges gracefully over the history and ideas of mathematics and physics, interwoven with an intimate diary of a young scientist as she embarks on her career.

inflationary cosmology: For a brief introduction to cosmology and inflation, see two articles by Stephen Battersby: "Introduction: Cosmology," *New Scientist* (September 4, 2006), online at http://www.newscientist.com/article/dn9988-introduction-cosmology.html, and "Best ever map of the early universe revealed," *New Scientist* (March 17, 2006), online at http://www.newscientist.com/article/dn8862-best-ever-map-of-the-early-universe-revealed

.html. The case for inflation remains controversial, however. Its strengths and weaknesses are explained in P. J. Steinhardt, "The inflation debate: Is the theory at the heart of modern cosmology deeply flawed?" *Scientific American* (April 2011), pp. 18–25.

16. Take It to the Limit

121 *Zeno*: The history and intellectual legacy of Zeno's paradoxes are discussed in J. Mazur, *Zeno's Paradox* (Plume, 2008).

122 *circles and pi*: For a delightfully opinionated and witty history of pi, see P. Beckmann, *A History of Pi* (St. Martin's Press, 1976).

126 *Archimedes used a similar strategy*: The PBS television series *Nova* ran a wonderful episode about Archimedes, infinity, and limits, called "Infinite Secrets." It originally aired on September 30, 2003. The program website (http://www.pbs.org/wgbh/nova/archimedes/) offers many online resources, including the program transcript and interactive demonstrations.

127 *method of exhaustion*: For readers wishing to see the mathematical details of Archimedes's method of exhaustion, Neal Carothers has used trigonometry (equivalent to the Pythagorean gymnastics that Archimedes relied on) to derive the perimeters of the inscribed and circumscribed polygons between which the circle is trapped; see http://personal.bgsu.edu/~carother/pi/Pi3a.html. Peter Alfeld's webpage "Archimedes and the computation of pi" features an interactive Java applet that lets you change the number of sides in the polygons; see http://www.math.utah.edu/~alfeld/Archimedes/Archimedes.html. The individual steps in Archimedes's original argument are of historical interest but you might find them disappointingly obscure; see http://itech.fgcu.edu/faculty/clindsey/mhf4404/archimedes/archimedes.html.

pi remains as elusive as ever: Anyone curious about the heroic computations of pi to immense numbers of digits should enjoy Richard Preston's profile of the Chudnovsky brothers. Entitled "The mountains of pi," this affectionate and surprisingly comical piece appeared in the March 2, 1992, issue of the *New Yorker*, and more recently as a chapter in R. Preston, *Panic in Level Four* (Random House, 2008).

numerical analysis: For a textbook introduction to the basics of numerical analysis, see W. H. Press, S. A. Teukolsky, W. T. Vetterling, and B. P. Flannery, *Numerical Recipes*, 3rd edition (Cambridge University Press, 2007).

17. Change We Can Believe In

131 *Every year about a million American students take calculus*: D. M. Bressoud, "The crisis of calculus," Mathematical Association of America (April 2007), available at http://www.maa.org/columns/launchings/launchings_04_07.html.

133 *Michael Jordan flying through the air*: For video clips of Michael Jordan's most spectacular dunks, see http://www.youtube.com/watch?v=H8M2NgjvicA.

134 *My high-school calculus teacher*: For a collection of Mr. Joffray's calculus problems, both classic and original, see S. Strogatz, *The Calculus of Friendship* (Princeton University Press, 2009).

137 *Snell's law*: Several articles, videos, and websites present the details of Snell's law and its derivation from Fermat's principle (which states that light takes the path of least time). For example, see M. Golomb, "Elementary proofs for the equivalence of Fermat's principle and Snell's law," *American Mathematical Monthly*, Vol. 71, No. 5 (May 1964), pp. 541–543, and http://en.wikibooks.org/wiki/Optics/Fermat%27s_Principle. Others provide historical accounts; see http://en.wikipedia.org/wiki/Snell%27s_law.

   Fermat's principle was an early forerunner to the more general principle of least action. For entertaining and deeply enlightening discussions of this principle, including its basis in quantum mechanics, see R. P. Feynman, R. B. Leighton, and M. Sands, "The principle of least action," *The Feynman Lectures on Physics*, Vol. 2, chapter 19 (Addison-Wesley, 1964), and R. Feynman, *QED* (Princeton University Press, 1988).

all possible paths: In a nutshell, Feynman's astonishing proposition is that nature actually does try all paths. But nearly all of the paths cancel one another out through a quantum analog of destructive interference — except for those very close to the classical path where the action is minimized (or, more precisely, made stationary). There the quantum interference becomes constructive, rendering those paths exceedingly more likely to be observed. This, in Feynman's account, is why nature obeys minimum principles. The key is that we live in the macroscopic world of everyday experience where the actions are colossal compared to Planck's constant. In that classical limit, quantum destructive interference becomes extremely strong and obliterates nearly everything else that could otherwise happen.

18. It Slices, It Dices

140 *oncology*: For more about the ways that integral calculus has been used to help cancer researchers, see D. Mackenzie, "Mathematical modeling of cancer," *SIAM News*, Vol. 37 (January/February 2004), and H. P. Greenspan, "Models for the growth of a solid tumor by diffusion," *Studies in Applied Mathematics* (December 1972), pp. 317–340.

141 *solid common to two identical cylinders*: The region common to two identical circular cylinders whose axes intersect at right angles is known variously as a Steinmetz solid or a bicylinder. For background, see http://mathworld.wolfram.com/SteinmetzSolid.html and http://en.wikipedia.org/wiki/Steinmetz_solid. The Wikipedia page also includes a very helpful computer animation that shows the Steinmetz solid emerging, ghostlike, from the intersecting cylinders. Its volume can be calculated straightforwardly but opaquely by modern techniques.

An ancient and much simpler solution was known to both Archimedes and Tsu Ch'ung-Chih. It uses nothing more than the method of slicing and a comparison between the areas of a square and a circle. For a marvelously clear exposition, see Martin Gardner's column "Mathematical games: Some puzzles based on checkerboards," *Scientific American*, Vol. 207 (November 1962), p. 164. And for Archimedes and Tsu Ch'ung-Chih, see: Archimedes, *The Method*, English translation by T. L. Heath (1912), reprinted by Dover (1953); and T. Kiang, "An old Chinese way of finding the volume of a sphere," *Mathematical Gazette*, Vol. 56 (May 1972), pp. 88–91.

Moreton Moore points out that the bicylinder also has applications in architecture: "The Romans and Normans, in using the barrel vault to span their buildings, were familiar with the geometry of intersecting cylinders where two such vaults crossed one another to form a cross vault." For this, as well as applications to crystallography, see M. Moore, "Symmetrical intersections of right circular cylinders," *Mathematical Gazette*, Vol. 58 (October 1974), pp. 181–185.

142 *Computer graphics*: Interactive demonstrations of the bicylinder and other problems in integral calculus are available online at the Wolfram Demonstrations Project (http://demonstrations.wolfram .com/). To play them, you'll need to download the free Mathematica Player (http://www.wolfram.com/products/player/), which will then allow you to explore hundreds of other interactive demonstrations in all parts of mathematics. The bicylinder demo is at http:// demonstrations.wolfram.com/IntersectingCylinders/.

Mamikon Mnatsakanian at Caltech has produced a series of animations that illustrate the Archimedean spirit and the power of slicing. My favorite is http://www.its.caltech.edu/~mamikon/ Sphere.html, which depicts a beautiful relationship among the volumes of a sphere and a certain double-cone and cylinder whose height and radius match those of the sphere. He also shows the same thing more physically by draining an imaginary volume of liquid from the cylinder and pouring it into the other two shapes; see http://www.its.caltech.edu/~mamikon/SphereWater.html.

Similarly elegant mechanical arguments in the service of math are given in M. Levi, *The Mathematical Mechanic* (Princeton University Press, 2009).

143 *Archimedes managed to do this*: For Archimedes's application of his mechanical method to the problem of finding the volume of the bicylinder, see T. L. Heath, ed., Proposition 15, *The Method of Archimedes, Recently Discovered by Heiberg* (Cosimo Classics, 2007), p. 48.

On p. 13 of the same volume, Archimedes confesses that he views his mechanical method as a means for discovering theorems rather than proving them: "certain things first became clear to me by a mechanical method, although they had to be demonstrated by geometry afterwards because their investigation by the said method did not furnish an actual demonstration. But it is of course easier, when we have previously acquired, by the method, some knowledge of the questions, to supply the proof than it is to find it without any previous knowledge."

For a popular account of Archimedes's work, see R. Netz and W. Noel, *The Archimedes Codex* (Da Capo Press, 2009).

19. All about *e*

147 *what is e, exactly*: For an introduction to all things *e* and exponential, see E. Maor, *e: The Story of a Number* (Princeton University Press, 1994). Readers with a background in calculus will enjoy the Chauvenet Prize–winning article by B. J. McCartin, "*e*: The master of all," *Mathematical Intelligencer*, Vol. 28, No. 2 (2006), pp. 10–21. A PDF version is available at http://mathdl.maa.org/images/upload_library/22/Chauvenet/mccartin.pdf.

151 *about 13.5 percent of the seats go to waste*: The expected packing fraction for couples sitting in a theater at random has been studied in other guises in the scientific literature. It first arose in organic chemistry—see P. J. Flory, "Intramolecular reaction between neighboring substituents of vinyl polymers," *Journal of the American Chemical Society*, Vol. 61 (1939), pp. 1518–1521. The magic number $1/e^2$ appears on the top right column of p. 1519. A more recent treatment relates this question to the random parking problem, a classic puzzle in probability theory and statistical physics; see W. H. Olson, "A Markov chain model for the kinetics of reactant isolation," *Journal of Applied Probability*, Vol. 15, No. 4 (1978), pp. 835–841. Computer scientists have tackled similar questions in their studies of "random greedy matching" algorithms for pairing neighboring nodes of networks; see M. Dyer and A. Frieze, "Randomized greedy matching," *Random Structures and Algorithms*, Vol. 2 (1991), pp. 29–45.

152 *how many people you should date:* The question of when to stop
dating and choose a mate has also been studied in various forms,
leading to such designations as the fiancée problem, the marriage
problem, the fussy-suitor problem, and the sultan's-dowry problem. But the most common term for it nowadays is the secretary
problem (the imagined scenario being that you're trying to hire the
best secretary from a given pool of candidates; you interview the
candidates one at a time and have to decide on the spot whether
to hire the person or say goodbye forever). For an introduction to
the mathematics and history of this wonderful puzzle, see http://
mathworld.wolfram.com/SultansDowryProblem.html and http://
en.wikipedia.org/wiki/Secretary_problem. For more details, see
T. S. Ferguson, "Who solved the secretary problem?" *Statistical
Science*, Vol. 4, No. 3 (1989), pp. 282–289. A clear exposition of
how to solve the problem is given at http://www.math.uah.edu/
stat/urn/Secretary.xhtml. For an introduction to the larger subject
of optimal stopping theory, see T. P. Hill, "Knowing when to stop:
How to gamble if you must — the mathematics of optimal stopping," *American Scientist*, Vol. 97 (2009), pp. 126–133.

20. Loves Me, Loves Me Not

155 *suppose Romeo is in love with Juliet:* For models of love affairs based
on differential equations, see section 5.3 in S. H. Strogatz, *Nonlinear
Dynamics and Chaos* (Perseus, 1994).

159 *"It is useful to solve differential equations"*: For Newton's anagram, see
p. vii in V. I. Arnold, *Geometrical Methods in the Theory of Ordinary
Differential Equations* (Springer, 1994).

159 *chaos:* Chaos in the three-body problem is discussed in I. Peterson,
Newton's Clock (W. H. Freeman, 1993).

160 *"made his head ache"*: For the quote about how the three-body problem made Newton's head ache, see D. Brewster, *Memoirs of the Life,
Writings, and Discoveries of Sir Isaac Newton* (Thomas Constable
and Company, 1855), Vol. 2, p. 158.

21. Step Into the Light

161 *vector calculus*: A great introduction to vector calculus and Maxwell's equations, and perhaps the best textbook I've ever read, is E. M. Purcell, *Electricity and Magnetism*, 2nd edition (Cambridge University Press, 2011). Another classic is H. M. Schey, *Div, Grad, Curl, and All That*, 4th edition (W. W. Norton and Company, 2005).

162 *Maxwell discovered what light is*: These words are being written during the 150th anniversary of Maxwell's 1861 paper "On physical lines of force." See, specifically, "Part III. The theory of molecular vortices applied to statical electricity," *Philosophical Magazine* (April and May, 1861), pp. 12–24, available at http://en.wikisource.org/wiki/On_Physical_Lines_of_Force and scanned from the original at http://www.vacuum-physics.com/Maxwell/maxwell_oplf.pdf.

The original paper is well worth a look. A high point occurs just below equation 137, where Maxwell—a sober man not prone to theatrics—couldn't resist italicizing the most revolutionary implication of his work: "The velocity of transverse undulations in our hypothetical medium, calculated from the electro-magnetic experiments of M. M. Kohlrausch and Weber, agrees so exactly with the velocity of light calculated from the optical experiments of M. Fizeau, that we can scarcely avoid the inference that *light consists in the transverse undulations of the same medium which is the cause of electric and magnetic phenomena*."

169 *airflow around a dragonfly as it hovered*: For Jane Wang's work on dragonfly flight, see Z. J. Wang, "Two dimensional mechanism for insect hovering," *Physical Review Letters*, Vol. 85, No. 10 (September 2000), pp. 2216–2219, and Z. J. Wang, "Dragonfly flight," *Physics Today*, Vol. 61, No. 10 (October 2008), p. 74. Her papers are also downloadable from http://dragonfly.tam.cornell.edu/insect.html. A video of dragonfly flight is at the bottom of http://ptonline.aip.org/journals/doc/PHTOAD-ft/vol_61/iss_10/74_1.shtml.

170 *How I wish I could have witnessed the moment*: Einstein also seems to have wished he were a fly on the wall in Maxwell's study. As he

wrote in 1940, "Imagine [Maxwell's] feelings when the differential equations he had formulated proved to him that electromagnetic fields spread in the form of polarized waves and at the speed of light! To few men in the world has such an experience been vouchsafed." See p. 489 in A. Einstein, "Considerations concerning the fundaments of theoretical physics," *Science*, Vol. 91 (May 24, 1940), pp. 487–492 (available online at http://www.scribd.com/doc/30217690/Albert-Einstein-Considerations-Concerning-the-Fundaments-of-Theoretical-Physics).

171 *the legacy of his conjuring with symbols*: Maxwell's equations are often portrayed as a triumph of pure reason, but Simon Schaffer, a historian of science at Cambridge, has argued they were driven as much by the technological challenge of the day: the problem of transmitting signals along undersea telegraph cables. See S. Schaffer, "The laird of physics," *Nature*, Vol. 471 (2011), pp. 289–291.

22. The New Normal

175 *the world is now teeming with data*: For the new world of data mining, see S. Baker, *The Numerati* (Houghton Mifflin Harcourt, 2008), and I. Ayres, *Super Crunchers* (Bantam, 2007).

Sports statisticians crunch the numbers: M. Lewis, *Moneyball* (W. W. Norton and Company, 2003).

"Learn some statistics": N. G. Mankiw, "A course load for the game of life," *New York Times* (September 4, 2010).

176 *"Take statistics"*: D. Brooks, "Harvard-bound? Chin up," *New York Times* (March 2, 2006).

central lessons of statistics: For informative introductions to statistics leavened by fine storytelling, see D. Salsburg, *The Lady Tasting Tea* (W. H. Freeman, 2001), and L. Mlodinow, *The Drunkard's Walk* (Pantheon, 2008).

Galton board: If you've never seen a Galton board in action, check out the demonstrations available on YouTube. One of the most dra-

matic of these videos makes use of sand rather than balls; see http://
www.youtube.com/watch?v=xDIyAOBa_yU.

177   *heights of adult men and women*: You can find your place on the
height distribution by using the online analyzer at http://www
.shortsupport.org/Research/analyzer.html. Based on data from
1994, it shows what fraction of the U.S. population is shorter
or taller than any given height. For more recent data, see M. A.
McDowell et al., "Anthropometric reference data for children
and adults: United States, 2003–2006," *National Health Statistics
Reports*, No. 10 (October 22, 2008), available online at http://www
.cdc.gov/nchs/data/nhsr/nhsr010.pdf.

178   *OkCupid*: OkCupid is the largest free dating site in the United
States, with 7 million active members as of summer 2011. Their
statisticians perform original analysis on the anonymized and ag-
gregated data they collect from their members and then post their
results and insights on their blog OkTrends (http://blog.okcupid
.com/index.php/about/). The height distributions are presented
in C. Rudder, "The big lies people tell in online dating," http://
blog.okcupid.com/index.php/the-biggest-lies-in-online-dating/.
Thanks to Christian Rudder for generously allowing me to adapt
the plots from his post.

180   *Power-law distributions*: Mark Newman gives a superb introduction to this topic in M. E. J. Newman, "Power laws, Pareto distributions and Zipf's law," *Contemporary Physics*, Vol. 46, No. 5
(2005), pp. 323–351 (available online at http://www-personal.
umich.edu/~mejn/courses/2006/cmplxsys899/powerlaws.pdf).
This article includes plots of word frequencies in *Moby-Dick*, the
magnitudes of earthquakes in California from 1910 to 1992, the
net worth of the 400 richest people in the United States in 2003,
and many of the other heavy-tailed distributions mentioned in this
chapter. An earlier but still excellent treatment of power laws is M.
Schroder, *Fractals, Chaos, Power Laws* (W. H. Freeman, 1991).

181   *2003 tax cuts*: I've borrowed this example from C. Seife, *Proofiness*
(Viking, 2010). The transcript of President Bush's speech is avail-

able at http://georgewbush-whitehouse.archives.gov/news/releases/ 2004/02/print/20040219-4.html. The figures used in the text are based on the analysis by FactCheck.org (a nonpartisan project of the Annenberg Public Policy Center of the University of Pennsylvania), available online at http://www.factcheck.org/here_ we_go_again_bush_exaggerates_tax.html, and this analysis published by the nonpartisan Tax Policy Center: W. G. Gale, P. Orszag, and I. Shapiro, "Distributional effects of the 2001 and 2003 tax cuts and their financing," http://www.taxpolicycenter.org/publications/ url.cfm?ID=411018.

fluctuations in stock prices: B. Mandelbrot and R. L. Hudson, *The (Mis)Behavior of Markets* (Basic Books, 2004); N. N. Taleb, *The Black Swan* (Random House, 2007).

182 *Fat, heavy, and long*: These three words aren't always used synonymously. When statisticians speak of a long tail, they mean something different than when business and technology people discuss it. For example, in Chris Anderson's *Wired* article "The long tail" from October 2004 (http://www.wired.com/wired/archive/12.10/ tail.html) and in his book of the same name, he's referring to the huge numbers of films, books, songs, and other works that are obscure to most of the population but that nonetheless have niche appeal and so survive online. In other words, for him, the long tail is the millions of little guys; for statisticians, the long tail is the very few big guys: the superwealthy, or the large earthquakes.

The difference is that Anderson switches the axes on his plots, which is somewhat like looking through the other end of the telescope. His convention is opposite that used by statisticians in their plots of cumulative distributions, but it has a long tradition going back to Vilfredo Pareto, an engineer and economist who studied the income distributions of European countries in the late 1800s. In a nutshell, Anderson and Pareto plot frequency as a function of rank, whereas Zipf and the statisticians plot rank as a function of frequency. The same information is shown either way, but with the axes flipped.

This leads to much confusion in the scientific literature. See http://www.hpl.hp.com/research/idl/papers/ranking/ranking.html

for Lada Adamic's tutorial sorting this out. Mark Newman also clarifies this point in his paper on power laws, mentioned above.

23. Chances Are

183 *probability theory*: For a good textbook treatment of conditional probability and Bayes's theorem, see S. M. Ross, *Introduction to Probability and Statistics for Engineers and Scientists*, 4th edition (Academic Press, 2009). For a history of Reverend Bayes and the controversy surrounding his approach to probabilistic inference, see S. B. McGrayne, *The Theory That Would Not Die* (Yale University Press, 2011).

183 *ailing plant*: The answer to part (a) of the ailing-plant problem is 59 percent. The answer to part (b) is 27/41, or approximately 65.85 percent. To derive these results, imagine 100 ailing plants and figure out how many of them (on average) get watered or not, and then how many of those go on to die or not, based on the information given. This question appears, though with slightly different numbers and wording, as problem 29 on p. 84 of Ross's text.

185 *mammogram*: The study of how doctors interpret mammogram results is described in G. Gigerenzer, *Calculated Risks* (Simon and Schuster, 2002), chapter 4.

187 *conditional-probability problems can still be perplexing*: For many entertaining anecdotes and insights about conditional probability and its real-world applications, as well as how it's misperceived, see J. A. Paulos, *Innumeracy* (Vintage, 1990), and L. Mlodinow, *The Drunkard's Walk* (Vintage, 2009).

187 *O.J. Simpson trial*: For more on the O.J. Simpson case and a discussion of wife battering in a larger context, see chapter 8 of Gigerenzer, *Calculated Risks*. The quotes pertaining to the O.J. Simpson trial and Alan Dershowitz's estimate of the rate at which battered women are murdered by their partners appeared in A. Dershowitz, *Reasonable Doubts* (Touchstone, 1997), pp. 101–104.

Probability theory was first correctly applied to the Simpson trial in 1995. The analysis given in this chapter is based on that proposed by the late I. J. Good in "When batterer turns murderer," *Nature*, Vol. 375 (1995), p. 541, and refined in "When batterer becomes murderer," *Nature*, Vol. 381 (1996), p. 481. Good phrased his analysis in terms of odds ratios and Bayes's theorem rather than in the more intuitive natural frequency approach used here and in Gigerenzer's book. (Incidentally, Good had an interesting career. In addition to his many contributions to probability theory and Bayesian statistics, he helped break the Nazi Enigma code during World War II and introduced the futuristic concept now known as the technological singularity.)

For an independent analysis that reaches essentially the same conclusion and that was also published in 1995, see J. F. Merz and J. P. Caulkins, "Propensity to abuse—propensity to murder?" *Chance*, Vol. 8, No. 2 (1995), p. 14. The slight differences between the two approaches are discussed in J. B. Garfield and L. Snell, "Teaching bits: A resource for teachers of statistics," *Journal of Statistics Education*, Vol. 3, No. 2 (1995), online at http://www.amstat.org/publications/jse/v3n2/resource.html.

188   *Dershowitz countered for the defense*: Here is how Dershowitz seems to have calculated that fewer than 1 in 2,500 batterers per year go on to murder their partners. On page 104 of his book *Reasonable Doubts*, he cites an estimate that in 1992, somewhere between 2.5 and 4 million women in the United States were battered by their husbands, boyfriends, and ex-boyfriends. In that same year, according to the FBI Uniform Crime Reports (http://www.fbi.gov/about-us/cjis/ucr/ucr), 913 women were murdered by their husbands, and 519 were killed by their boyfriends or ex-boyfriends. Dividing the total of 1,432 homicides by 2.5 million beatings yields 1 murder per 1,746 beatings, whereas using the higher estimate of 4 million beatings per year yields 1 murder per 2,793 beatings. Dershowitz apparently chose 2,500 as a round number in between these extremes.

What's unclear is what proportion of the murdered women had previously been beaten by these men. It seems that Dershowitz was assuming that nearly all the homicide victims had earlier been beaten, presumably to make the point that even when the rate is overestimated in this way, it's still "infinitesimal."

A few years after the verdict was handed down in the Simpson case, Dershowitz and the mathematician John Allen Paulos engaged in a heated exchange via letters to the editor of the *New York Times*. The issue was whether evidence of a history of spousal abuse should be regarded as relevant to a murder trial in light of probabilistic arguments similar to those discussed here. See A. Dershowitz, "The numbers game," *New York Times* (May 30, 1999), archived at http://www.nytimes.com/1999/05/30/books/l-the-numbers-game -789356.html, and J. A. Paulos, "Once upon a number," *New York Times* (June 27, 1999), http://www.nytimes.com/1999/06/27/ books/l-once-upon-a-number-224537.html.

expect 3 more of these women, on average, to be killed by someone else: According to the FBI Uniform Crime Reports, 4,936 women were murdered in 1992. Of these murder victims, 1,432 (about 29 percent) were killed by their husbands or boyfriends. The remaining 3,504 were killed by somebody else. Therefore, considering that the total population of women in the United States at that time was about 125 million, the rate at which women were murdered by someone other than their partners was 3,504 divided by 125,000,000, or 1 murder per 35,673 women per year.

Let's assume that this rate of murder by nonpartners was the same for all women, battered or not. Then in our hypothetical sample of 100,000 battered women, we'd expect about 100,000 divided by 35,673, or 2.8, women to be killed by someone other than a partner. Rounding 2.8 to 3, we obtain the estimate given in the text.

24. Untangling the Web

191   *searching the Web*: For an introduction to Web search and link analysis, see D. Easley and J. Kleinberg, *Networks, Crowds, and Markets* (Cambridge University Press, 2010), chapter 14. Their elegant exposition has inspired my treatment here. For a popular account of the history of Internet search, including stories about the key characters and companies, see J. Battelle, *The Search* (Portfolio Hardcover, 2005). The early development of link analysis, for readers comfortable with linear algebra, is summarized in S. Robinson,

"The ongoing search for efficient Web search algorithms," *SIAM News*, Vol. 37, No. 9 (2004).

grasshopper: For anyone confused by my use of the word "grasshopper," it is an affectionate nickname for a student who has much to learn from a Zen master. In the television series *Kung Fu*, on many of the occasions when the blind monk Po imparts wisdom to his student Caine, he calls him grasshopper, harking back to their first lesson, a scene depicted in the 1972 pilot film (and online at http://www.youtube.com/watch?v=WCyJRXvPNRo):

> Master Po: Close your eyes. What do you hear?
> Young Caine: I hear the water. I hear the birds.
> Po: Do you hear your own heartbeat?
> Caine: No.
> Po: Do you hear the grasshopper which is at your feet?
> Caine: Old man, how is it that you hear these things?
> Po: Young man, how is it that you do not?

circular reasoning: The recognition of the circularity problem for ranking webpages and its solution via linear algebra grew out of two lines of research published in 1998. One was by my Cornell colleague Jon Kleinberg, then working as a visiting scientist at IBM Almaden Research Center. For his seminal paper on the "hubs and authorities" algorithm (an alternative form of link analysis that appeared slightly earlier than Google's PageRank algorithm), see J. Kleinberg, "Authoritative sources in a hyperlinked environment," *Proceedings of the Ninth Annual ACM-SIAM Symposium on Discrete Algorithms* (1998).

The other line of research was by Google cofounders Larry Page and Sergey Brin. Their PageRank method was originally motivated by thinking about the proportion of time a random surfer would spend at each page on the Web — a process with a different description but one that leads to the same way of resolving the circular definition. The foundational paper on PageRank is S. Brin and L. Page, "The anatomy of a large-scale hypertextual Web search engine," *Proceedings of the Seventh International World Wide Web Conference* (1998), pp. 107–117.

As so often happens in science, strikingly similar precursors of these ideas had already been discovered in other fields. For this pre-history of PageRank in bibliometrics, psychology, sociology, and econometrics, see M. Franceschet, "PageRank: Standing on the shoulders of giants," *Communications of the ACM*, Vol. 54, No. 6 (2011), available at http://arxiv.org/abs/1002.2858; and S. Vigna, "Spectral ranking," http://arxiv.org/abs/0912.0238.

linear algebra: For anyone seeking an introduction to linear algebra and its applications, Gil Strang's books and online video lectures would be a fine place to start: G. Strang, *Introduction to Linear Algebra*, 4th edition (Wellesley-Cambridge Press, 2009), and http://web.mit.edu/18.06/www/videos.html.

linear algebra has the tools you need: Some of the most impressive applications of linear algebra rely on the techniques of singular value decomposition and principal component analysis. See D. James, M. Lachance, and J. Remski, "Singular vectors' subtle se-crets," *College Mathematics Journal*, Vol. 42, No. 2 (March 2011), pp. 86–95.

PageRank algorithm: According to Google, the term "PageRank" refers to Larry Page, not webpage. See http://web.archive.org/web/20090424093934/http://www.google.com/press/funfacts.html.

classify human faces: The idea here is that any human face can be expressed as a combination of a small number of fundamental face ingredients, or eigenfaces. This application of linear algebra to face recognition and classification was pioneered by L. Sirovich and M. Kirby, "Low-dimensional procedure for the characterization of hu-man faces," *Journal of the Optical Society of America A*, Vol. 4 (1987), pp. 519–524, and further developed by M. Turk and A. Pentland, "Eigenfaces for recognition," *Journal of Cognitive Neuroscience*, Vol. 3 (1991), pp. 71–86, which is also available online at http://cse.seu.edu.cn/people/xgeng/files/under/turk91eigenfaceForRecognition.pdf.

For a comprehensive list of scholarly papers in this area, see the Face Recognition Homepage (http://www.face-rec.org/interesting-papers/).

voting patterns of Supreme Court justices: L. Sirovich, "A pattern analysis of the second Rehnquist U.S. Supreme Court," *Proceedings of the National Academy of Sciences*, Vol. 100, No. 13 (2003), pp. 7432–7437. For a journalistic account of this work, see N. Wade, "A mathematician crunches the Supreme Court's numbers," *New York Times* (June 24, 2003). For a discussion aimed at legal scholars by a mathematician and now law professor, see P. H. Edelman, "The dimension of the Supreme Court," *Constitutional Commentary*, Vol. 20, No. 3 (2003), pp. 557–570.

Netflix Prize: For the story of the Netflix Prize, with amusing details about its early contestants and the importance of the movie *Napoleon Dynamite*, see C. Thompson, "If you liked this, you're sure to love that—Winning the Netflix prize," *New York Times Magazine* (November 23, 2008). The prize was won in September 2009, three years after the contest began; see S. Lohr, "A $1 million research bargain for Netflix, and maybe a model for others," *New York Times* (September 22, 2009). The application of the singular value decomposition to the Netflix Prize is discussed in B. Cipra, "Blockbuster algorithm," *SIAM News*, Vol. 42, No. 4 (2009).

193 *skipping some details*: For simplicity, I have presented only the most basic version of the PageRank algorithm. To handle networks with certain common structural features, PageRank needs to be modified. Suppose, for example, that the network has some pages that point to others but that have none pointing back to them. During the update process, those pages will lose their PageRank, as if leaking or hemorrhaging it. They give it to others but it's never replenished. So they'll all end up with PageRanks of zero and will therefore be indistinguishable in that respect.

At the other extreme, consider networks in which some pages, or groups of pages, hoard PageRank by being clubby, by never linking back out to anyone else. Such pages tend to act as sinks for PageRank.

To overcome these and other effects, Brin and Page modified their algorithm as follows: After each step in the update process, all the current PageRanks are scaled down by a constant factor, so the total is less than 1. Whatever is left over is then evenly distributed to all the nodes in the network, as if being rained down on everyone. It's the ultimate egalitarian act, spreading the PageRank to the neediest nodes. Joe the Plumber would not be happy.

For a deeper look at the mathematics of PageRank, with interactive explorations, see E. Aghapour, T. P. Chartier, A. N. Langville, and K. E. Pedings, "Google PageRank: The mathematics of Google" (http://www.whydomath.org/node/google/index.html). A comprehensive yet accessible book-length treatment is A. N. Langville and C. D. Meyer, *Google's PageRank and Beyond* (Princeton University Press, 2006).

25. The Loneliest Numbers

201 *one is the loneliest number*: Harry Nilsson wrote the song "One." Three Dog Night's cover of it became a hit, reaching number 5 on the *Billboard* Hot 100, and Aimee Mann has a terrific version of it that can be heard in the movie *Magnolia*.

The Solitude of Prime Numbers: P. Giordano, *The Solitude of Prime Numbers* (Pamela Dorman Books/Viking Penguin, 2010). The passage excerpted here appears on pp. 111–112.

202 *number theory*: For popular introductions to number theory, and the mysteries of prime numbers in particular, the most difficult choice is where to begin. There are at least three excellent books to choose from. All appeared at around the same time, and all centered on the Riemann hypothesis, widely regarded as the greatest unsolved problem in mathematics. For some of the mathematical details, along with the early history of the Riemann hypothesis, I'd recommend J. Derbyshire, *Prime Obsession* (Joseph Henry Press, 2003). For more emphasis on the latest developments but still at a very accessible level, see D. Rockmore, *Stalking the Riemann Hypothesis* (Pantheon, 2005), and M. du Sautoy, *The Music of the Primes* (HarperCollins, 2003).

203 *encryption algorithms*: For the use of number theory in cryptography, see M. Gardner, *Penrose Tiles to Trapdoor Ciphers* (Mathematical Association of America, 1997), chapters 13 and 14. The first of these chapters reprints Gardner's famous column from the August

1977 issue of *Scientific American* in which he told the public about the virtually unbreakable RSA cryptosystem. The second chapter describes the "intense furor" it aroused within the National Security Agency. For more recent developments, see chapter 10 of du Sautoy, *The Music of the Primes*.

209 *prime number theorem*: Along with the books by Derbyshire, Rockmore, and du Sautoy mentioned above, there are many online sources of information about the prime number theorem, such as Chris K. Caldwell's page "How many primes are there?" (http://primes.utm.edu/howmany.shtml), the MathWorld page "Prime number theorem" (http://mathworld.wolfram.com/PrimeNumberTheorem.html), and the Wikipedia page "Prime number theorem" (http://en.wikipedia.org/wiki/Prime_number_theorem).

Carl Friedrich Gauss: The story of how Gauss noticed the prime number theorem at age fifteen is told on pp. 53–54 of Derbyshire, *Prime Obsession*, and in greater detail by L. J. Goldstein, "A history of the prime number theorem," *American Mathematical Monthly*, Vol. 80, No. 6 (1973), pp. 599–615. Gauss did not prove the theorem but guessed it by poring over tables of prime numbers that he had computed—by hand—for his own amusement. The first proofs were published in 1896, about a century later, by Jacques Hadamard and Charles de la Vallée Poussin, each of whom had been working independently on the problem.

twin primes apparently continue to exist: How can twin primes exist at large N, in light of the prime number theorem? The theorem says only that the *average* gap is $\ln N$. But there are fluctuations about this average, and since there are infinitely many primes, some of them are bound to get lucky and beat the odds. In other words, even though most won't find a neighboring prime much closer than $\ln N$ away, some of them will.

For readers who want to see some of the math governing "very small gaps between primes" beautifully explained in a concise way, see Andrew Granville's article on analytic number theory in

T. Gowers, *The Princeton Companion to Mathematics* (Princeton University Press, 2008), pp. 332–348, especially p. 343.

Also, there's a nice online article by Terry Tao that gives a lot of insight into twin primes—specifically, how they're distributed and why mathematicians believe there are infinitely many of them—and then wades into deeper waters to explain the proof of his celebrated theorem (with Ben Green) that the primes contain arbitrarily long arithmetic progressions. See T. Tao, "Structure and randomness in the prime numbers," http://terrytao.wordpress.com/2008/01/07/ams-lecture-structure-and-randomness-in-the-prime-numbers/.

For further details and background information about twin primes, see http://en.wikipedia.org/wiki/Twin_prime and http://mathworld.wolfram.com/TwinPrimeConjecture.html.

another prime couple nearby: I'm just kidding around here and not trying to make a serious point about the spacing between consecutive pairs of twin primes. Maybe somewhere out there, far down the number line, two sets of twins happen to be extremely close together. For an introduction to such questions, see I. Peterson, "Prime twins" (June 4, 2001), http://www.maa.org/mathland/mathtrek_6_4_01.html.

In any case, the metaphor of awkward couples as twin primes has not been lost on Hollywood. For light entertainment, you might want to rent a movie called *The Mirror Has Two Faces*, a Barbra Streisand vehicle costarring Jeff Bridges. He's a handsome but socially clueless math professor. She's a professor in the English literature department, a plucky, energetic, but homely woman (or at least that's how we're supposed to see her) who lives with her mother and gorgeous sister. Eventually, the two professors manage to get together for their first date. When their dinner conversation drifts to the topic of dancing (which embarrasses him), he changes the subject abruptly to twin primes. She gets the idea immediately and asks, "What would happen if you counted past a million? Would there still be pairs like that?" He practically falls off his chair and says, "I can't believe you thought of that! That is exactly what is yet to be proven in the twin prime conjecture." Later in the movie, when they start falling in love, she gives him a birthday present of cuff links with prime numbers on them.

26. Group Think

212   *mattress math*: The mattress group is technically known as the Klein four-group. It's one of the simplest in a gigantic zoo of possibilities. Mathematicians have been analyzing groups and classifying their structures for about 200 years. For an engaging account of group theory and the more recent quest to classify all the finite simple groups, see M. du Sautoy, *Symmetry* (Harper, 2008).

group theory: Two recent books inspired this chapter: N. Carter, *Visual Group Theory* (Mathematical Association of America, 2009); and B. Hayes, *Group Theory in the Bedroom* (Hill and Wang, 2008). Carter introduces the basics of group theory gently and pictorially. He also touches on its connections to Rubik's cube, contra dancing and square dancing, crystals, chemistry, art, and architecture. An earlier version of Hayes's mattress-flipping article appeared in *American Scientist*, Vol. 93, No. 5 (September/October 2005), p. 395, available online at http://www.americanscientist.org/issues/pub/group-theory-in-the-bedroom.

Readers interested in seeing a definition of what a "group" is should consult any of the authoritative online references or standard textbooks on the subject. A good place to start is the MathWorld page http://mathworld.wolfram.com/topics/GroupTheory.html or the Wikipedia page http://en.wikipedia.org/wiki/Group_(mathematics). The treatment I've given here emphasizes symmetry groups rather than groups in the most general sense.

chaotic counterparts: Michael Field and Martin Golubitsky have studied the interplay between group theory and nonlinear dynamics. In the course of their investigations, they've generated stunning computer graphics of symmetric chaos, many of which can be found on Mike Field's webpage (http://www.math.uh.edu/%7Emike/ag/art.html). For the art, science, and mathematics of this topic, see M. Field and M. Golubitsky, *Symmetry in Chaos*, 2nd edition (Society for Industrial and Applied Mathematics, 2009).

217 *the diagram demonstrates that HR = V*: A word about some potentially confusing notation used throughout this chapter: In equations like *HR* = *V*, the *H* was written on the left to indicate that it's
the transformation being performed first. Carter uses this notation
for functional composition in his book, but the reader should be
aware that many mathematicians use the opposite convention, plac-
ing the *H* on the right.

Feynman got a draft deferment: For the anecdote about Feynman
and the psychiatrist, see R. P. Feynman, *"Surely You're Joking, Mr.
Feynman!"* (W. W. Norton and Company, 1985), p. 158, and J.
Gleick, *Genius* (Random House, 1993), p. 223.

27. Twist and Shout

219 *Möbius strips*: Art, limericks, patents, parlor tricks, and serious
math—you name it, and if it has anything to do with Möbius
strips, it's probably in Cliff Pickover's jovial book *The Möbius Strip*
(Basic Books, 2006). An earlier generation first learned about some
of these wonders in M. Gardner, "The world of the Möbius strip:
Endless, edgeless, and one-sided," *Scientific American*, Vol. 219, No.
6 (December 1968).

220 *fun activities that a six-year-old can do*: For step-by-step instructions,
with photographs, of some of the activities described in this chapter,
see "How to explore a Möbius strip" at http://www.wiki-how.com/
Explore-a-Möbius-Strip. Julian Fleron gives many other ideas—
Möbius garlands, hearts, paper clip stars—in "Recycling Möbius,"
http://artofmathematics.wsc.ma.edu/sculpture/workinprogress/
Mobius1206.pdf.

For further fun with paper models, see S. Barr's classic book
Experiments in Topology (Crowell, 1964).

topology: The basics of topology are expertly explained in R. Courant
and H. Robbins (revised by I. Stewart), *What Is Mathematics?* 2nd
edition (Oxford University Press, 1996), chapter 5. For a playful
survey, see M. Gardner, *The Colossal Book of Mathematics* (W. W.
Norton and Company, 2001). He discusses Klein bottles, knots,
linked doughnuts, and other delights of recreational topology in

part 5, chapters 18–20. A very good contemporary treatment is
D. S. Richeson, *Euler's Gem* (Princeton University Press, 2008).
Richeson presents a history and celebration of topology and an introduction to its main concepts, using Euler's polyhedron formula
as a centerpiece. At a much higher level, but still within comfortable
reach of people with a college math background, see the chapters
on algebraic topology and differential topology in T. Gowers, *The
Princeton Companion to Mathematics* (Princeton University Press,
2008), pp. 383–408.

221 *the intrinsic loopiness of a circle and a square*: Given that a circle and
square are topologically equivalent curves, you might be wondering
what kinds of curves would be topologically different. The simplest
example is a line segment. To prove this, observe that if you travel
in one direction around a circle, a square, or any other kind of loop,
you'll always return to your starting point, but that's not true for
traveling on a line segment. Since this property is left unchanged by
all transformations that preserve an object's topology (namely, continuous deformations whose inverse is also continuous), and since
this property differs between loops and segments, we can conclude
that loops and segments are topologically different.

223 *Vi Hart*: Vi's videos discussed in this chapter, "Möbius music box"
and "Möbius story: Wind and Mr. Ug," can be found on YouTube
and also at http://vihart.com/musicbox/ and http://vihart.com/
blog/mobius-story/. For more of her ingenious and fun-loving excursions into mathematical food, doodling, balloons, beadwork,
and music boxes, see her website at http://vihart.com/everything/.
She was profiled in K. Chang, "Bending and stretching classroom lessons to make math inspire," *New York Times* (January 17,
2011), available online at http://www.nytimes.com/2011/01/18/
science/18prof.html.

226 *artists have likewise drawn inspiration*: To see images of the Möbius
artwork by Maurits Escher, Max Bill, and Keizo Ushio, search the
Web using the artist's name and "Möbius" as search terms. Ivars
Peterson has written about the use of Möbius strips in literature, art,

architecture, and sculpture, with photographs and explanations, at his Mathematical Tourist blog: http://mathtourist.blogspot.com/search/label/Moebius%20Strips.

National Library of Kazakhstan: The library is currently under construction. For its design concept and intriguing images of what it will look like, go to the website for the architectural firm BIG (Bjarke Ingels Group), http://www.big.dk/. Click on the icon for ANL, Astana National Library; it appears in the 2009 column (the fourth column from the right) when the projects are arranged in their default chronological order. The site contains forty-one slides of the library's internal and external structure, museum circulation, thermal exposure, and so on, all of which are unusual because of the building's Möbius layout. For a profile of Bjarke Ingels and his practice, see G. Williams, "Open source architect: Meet the maestro of 'hedonistic sustainability,'" http://www.wired.co.uk/magazine/archive/2011/07/features/open-source-architect.

227 *Möbius patents*: Some of these are discussed in Pickover, *The Möbius Strip*. You can find hundreds of others by searching for "Möbius strip" in Google Patents.

bagel: If you want to try cutting a bagel this way, George Hart explains his technique at his website http://www.georgehart.com/bagel/bagel.html. Or you can see a computer animation by Bill Giles here: http://www.youtube.com/watch?v=hYXnZ8-ux80. If you prefer to watch it happening in real time, check out a video by UltraNurd called "Möbius Bagel" (http://www.youtube.com/watch?v=Zu5z1BCC70s). But strictly speaking, this should not be called a Möbius bagel—a point of confusion among many people who have written about or copied George's work. The surface on which the cream cheese is spread is *not* equivalent to a Möbius strip because it has two half twists in it, not one, and the resulting surface is two-sided, not one-sided. Furthermore, a true Möbius bagel would remain in one piece, not two, after being cut in half. For a demonstration of how to cut a bagel in this genuine Möbius fashion, see http://www.youtube.com/watch?v=l6Vuh16r8o8.

28. Think Globally

229   *a vision of the world as flat*: By referring to plane geometry as flat-earth geometry, I might seem to be disparaging the subject, but that's not my intent. The tactic of locally approximating a curved shape by a flat one has often turned out to be a useful simplification in many parts of mathematics and physics, from calculus to relativity theory. Plane geometry is the first instance of this great idea.

Nor do I mean to suggest that all the ancients thought the world was flat. For an engaging account of Eratosthenes's measurement of the distance around the globe, see N. Nicastro, *Circumference* (St. Martin's Press, 2008). For a more contemporary approach that you might like to try on your own, Robert Vanderbei at Princeton University recently gave a presentation to his daughter's high-school geometry class in which he used a photograph of a sunset to show that the Earth is not flat and to estimate its diameter. His slides are posted at http://orfe.princeton.edu/~rvdb/tex/sunset/34-39 .OPN.1108twoup.pdf.

differential geometry: A superb introduction to modern geometry was coauthored by David Hilbert, one of the greatest mathematicians of the twentieth century. This classic, originally published in 1952, has been reissued as D. Hilbert and S. Cohn-Vossen, *Geometry and the Imagination* (American Mathematical Society, 1999). Several good textbooks and online courses in differential geometry are listed on the Wikipedia page http://en.wikipedia.org/ wiki/Differential_geometry.

230   *most direct route*: For an interactive online demonstration that lets you plot the shortest route between any two points on the surface of the Earth, see http://demonstrations.wolfram.com/ GreatCirclesOnMercatorsChart/. (You'll need to download the free Mathematica Player, which will then allow you to explore hundreds of other interactive demonstrations in all parts of mathematics.)

234   *Konrad Polthier*: Excerpts from a number of Polthier's fascinating educational videos about mathematical topics can be found online at http://page.mi.fu-berlin.de/polthier/video/Geodesics/Scenes

.html. Award-winning videos by Polthier and his colleagues appear in the VideoMath Festival collection (http://page.mi.fu-berlin.de/ polthier/Events/VideoMath/index.html), available as a DVD from Springer-Verlag. For more details, see G. Glaeser and K. Polthier, *A Mathematical Picture Book* (Springer, 2012). The images shown in the text are from the DVD *Touching Soap Films* (Springer, 1995), by Andreas Arnez, Konrad Polthier, Martin Steffens, and Christian Teitzel.

236 *shortest path through a network*: The classic algorithm for short-est-path problems on networks was created by Edsger Dijkstra. For an introduction, see http://en.wikipedia.org/wiki/Dijkstra's_ algorithm. Steven Skiena has posted an instructive animation of Dijkstra's algorithm at http://www.cs.sunysb.edu/~skiena/ combinatorica/animations/dijkstra.html.

Nature can solve certain shortest-path problems by decentral-ized processes akin to analog computation. For chemical waves that solve mazes, see O. Steinbock, A. Toth, and K. Showalter, "Navigating complex labyrinths: Optimal paths from chemical waves," *Science*, Vol. 267 (1995), p. 868. Not to be outdone, slime molds can solve them too: T. Nakagaki, H. Yamada, and A. Toth, "Maze-solving by an amoeboid organism," *Nature*, Vol. 407 (2000), p. 470. This slimy creature can even make networks as efficient as the Tokyo rail system: A. Tero et al., "Rules for biologically inspired adaptive network design," *Science*, Vol. 327 (2010), p. 439.

telling the story of your life in six words: Delightful examples of six-word memoirs are given at http://www.smithmag.net/sixwords/ and http://en.wikipedia.org/wiki/Six-Word_Memoirs.

29. Analyze This!

238 *analysis*: Analysis grew out of the need to shore up the logical foun-dations of calculus. William Dunham traces this story through the works of eleven masters, from Newton to Lebesgue, in W. Dunham, *The Calculus Gallery* (Princeton University Press, 2005). The book contains explicit mathematics presented accessibly for readers having a college-level background. A textbook in a similar spirit

is D. Bressoud, *A Radical Approach to Real Analysis*, 2nd edition (Mathematical Association of America, 2006). For a more comprehensive historical account, see C. B. Boyer, *The History of the Calculus and Its Conceptual Development* (Dover, 1959).

vacillating forever: Grandi's series $1 - 1 + 1 - 1 + 1 - 1 + \cdots$ is discussed in a meticulously sourced Wikipedia article about its history, with links to further threads about its mathematical status and its role in math education. All these may be reached from the main page "Grandi's series": http://en.wikipedia.org/wiki/Grandi's_series.

243 *Riemann rearrangement theorem*: For a clear exposition of the Riemann rearrangement theorem, see Dunham, *The Calculus Gallery*, pp. 112–115.

244 *Strange, yes. Sick, yes*: The alternating harmonic series is conditionally convergent, meaning it's convergent but not absolutely convergent (the sum of the absolute values of its terms do *not* converge). For a series like that, you can reorder the sum to get *any* real number. That's the shocking implication of the Riemann rearrangement theorem. It shows that a convergent sum can violate our intuitive expectations if it does not converge *absolutely*.

In the much-better-behaved case of an absolutely convergent series, all rearrangements of the series converge to the same value. That's wonderfully convenient. It means that an absolutely convergent series behaves like a finite sum. In particular, it obeys the commutative law of addition. You can rearrange the terms any way you want without changing the answer. For more on absolute convergence, see http://mathworld.wolfram.com/AbsoluteConvergence.html and http://en.wikipedia.org/wiki/Absolute_convergence.

Fourier analysis: Tom Körner's extraordinary book *Fourier Analysis* (Cambridge University Press, 1989) is a self-described "shop window" of the ideas, techniques, applications, and history of Fourier analysis. The level of mathematical rigor is high, yet the book is witty, elegant, and pleasantly quirky. For an introduction to Fourier's work and its connection to music, see M. Kline, *Mathematics in Western Culture* (Oxford University Press, 1974), chapter 19.

247 *Gibbs phenomenon*: The Gibbs phenomenon and its tortuous history is reviewed by E. Hewitt and R. E. Hewitt, "The Gibbs-Wilbraham phenomenon: An episode in Fourier analysis," *Archive for the History of Exact Sciences*, Vol. 21 (1979), pp. 129–160.

digital photographs and on MRI scans: The Gibbs phenomenon can affect MPEG and JPEG compression of digital video: http://www.doc.ic.ac.uk/~nd/surprise_96/journal/vol4/sab/report.html. When it appears in MRI scans, the Gibbs phenomenon is known as truncation or Gibbs ringing; see http://www.mr-tip.com/serv1.php?type=art&sub=Gibbs%20Artifact. For methods to handle these artifacts, see T. B. Smith and K. S. Nayak, "MRI artifacts and correction strategies," *Imaging Medicine*, Vol. 2, No. 4 (2010), pp. 445–457, online at http://mrel.usc.edu/pdf/Smith_IM_2010.pdf.

pinpointed what causes Gibbs artifacts: The analysts of the 1800s identified the underlying mathematical cause of the Gibbs phenomenon. For functions (or, nowadays, images) displaying sharp edges or other mild types of jump discontinuities, the partial sums of the sine waves were proven to converge pointwise but not uniformly to the original function. Pointwise convergence means that at any *particular* point x, the partial sums get arbitrarily close to the original function as more terms are added. So in that sense, the series does converge, as one would hope. The catch is that some points are much more finicky than others. The Gibbs phenomenon occurs near the worst of those points — the edges in the original function.

For example, consider the sawtooth wave discussed in this chapter. As x gets closer to the edge of a sawtooth, it takes more and more terms in the Fourier series to reach a given level of approximation. That's what we mean by saying the convergence is not uniform. It occurs at different rates for different x.

In this case, the non-uniformity of the convergence is attributable to the pathologies of the alternating harmonic series, whose terms appear as Fourier coefficients for the sawtooth wave. As discussed above, the alternating harmonic series converges, but only because of the massive cancellation caused by its alternating mix of positive and negative terms. If all its terms were made positive by taking their absolute values, the series would diverge—the sum would approach infinity. That's why the alternating harmonic series is said to converge *conditionally*, not *absolutely*. This precarious form of convergence then infects the associated Fourier series and renders it non-uniformly convergent, leading to the Gibbs phenomenon and its mocking upraised finger near the edge.

In contrast, in the much better case where the series of Fourier coefficients *is* absolutely convergent, the associated Fourier series converges uniformly to the original function. Then the Gibbs phenomenon doesn't occur. For more details, see http://mathworld .wolfram.com/GibbsPhenomenon.html and http://en.wikipedia .org/wiki/Gibbs_phenomenon.

The bottom line is that the analysts taught us to be wary of conditionally convergent series. Convergence is good, but not good enough. For an infinite series to behave like a finite sum in all respects, it needs much tighter constraints than conditional convergence can provide. Insisting on absolute convergence yields the behavior we'd expect intuitively, both for the series itself and for its associated Fourier series.

30. The Hilbert Hotel

250 *Georg Cantor's*: For more about Cantor, including the mathematical, philosophical, and theological controversies surrounding his work, see J. W. Dauben, *Georg Cantor* (Princeton University Press, 1990).

set theory: If you haven't read it yet, I recommend the surprise bestseller *Logicomix*, a brilliantly creative graphic novel about set theory, logic, infinity, madness, and the quest for mathematical truth: A. Doxiadis and C. H. Papadimitriou, *Logicomix* (Bloomsbury, 2009). It stars Bertrand Russell, but Cantor, Hilbert, Poincaré, and many others make memorable appearances.

David Hilbert: The classic biography of David Hilbert is a moving and nontechnical account of his life, his work, and his times:

C. Reid, *Hilbert* (Springer, 1996). Hilbert's contributions to mathematics are too numerous to list here, but perhaps his greatest is his collection of twenty-three problems—all of which were unsolved when he proposed them—that he thought would shape the course of mathematics in the twentieth century. For the ongoing story and significance of these Hilbert problems and the people who solved some of them, see B. H. Yandell, *The Honors Class* (A K Peters, 2002). Several of the problems still remain open.

Hilbert Hotel: Hilbert's parable of the infinite hotel is mentioned in George Gamow's evergreen masterpiece *One Two Three . . . Infinity* (Dover, 1988), p. 17. Gamow also does a good job of explaining countable and uncountable sets and related ideas about infinity.

The comedic and dramatic possibilities of the Hilbert Hotel have often been explored by writers of mathematical fiction. For example, see S. Lem, "The extraordinary hotel or the thousand and first journey of Ion the Quiet," reprinted in *Imaginary Numbers*, edited by W. Frucht (Wiley, 1999), and I. Stewart, *Professor Stewart's Cabinet of Mathematical Curiosities* (Basic Books, 2009). A children's book on the same theme is I. Ekeland, *The Cat in Numberland* (Cricket Books, 2006).

256   *any* other *digit between 1 and 8*: A tiny finesse occurred in the argument for the uncountability of the real numbers when I required that the diagonal digits were to be replaced by digits between 1 and 8. This wasn't essential. But I wanted to avoid using 0 and 9 to sidestep any fussiness caused by the fact that some real numbers have two decimal representations. For example, .200000 . . . equals .199999 . . . Thus, if we hadn't excluded the use of 0s and 9s as replacement digits, it's conceivable the diagonal argument could have inadvertently produced a number already on the list (and that would have ruined the proof). By my forbidding the use of 0 and 9, we didn't have to worry about this annoyance.

infinity beyond infinity: For a more mathematical but still very readable discussion of infinity (and many of the other ideas discussed in this book), see J. C. Stillwell, *Yearning for the Impossible* (A K Peters, 2006). Readers wishing to go deeper into infinity might enjoy Terry Tao's blog post about self-defeating objects, http://terrytao.wordpress.com/2009/11/05/the-no-self-defeating-object-argument/. In a very accessible way, he presents and elucidates a lot of fundamental arguments about infinity that arise in set theory, philosophy, physics, computer science, game theory, and logic. For a survey of the foundational issues raised by these sorts of ideas, see also J. C. Stillwell, *Roads to Infinity* (A K Peters, 2010).

Credits

p. 3: "Sesame Workshop"®, "Sesame Street"® and associated characters, trademarks, and design elements are owned and licensed by Sesame Workshop. © 2011 Sesame Workshop. All Rights Reserved.

p. 41: Mark H. Anbinder

pp. 56, 57: Simon Tatham

p. 75: ABC Photo Archives / ABC via Getty Images

p. 78: Photograph © 2011 WET. All Rights Reserved.

p. 101: Garry Jenkins

p. 117: L. Clarke / Corbis

p. 118: Corbis

p. 119: Photodisc / Getty Images

p. 133: Manny Millan / Getty Images

pp. 141, 142: Paul Bourke

p. 151: J. R. Eyerman / Getty Images

p. 165: Alchemy / Alamy

p. 178: Christian Rudder / OkCupid

pp. 179, 180: M.E.J. Newman

p. 212: *Study for Alhambra Stars* (2000) / Mike Field

pp. 224, 225: Vi Hart

p. 226: BIG — Bjarke Ingels Group

p. 227: © George W. Hart

pp. 234, 235: Andreas Arnez, Konrad Polthier, Martin Steffens, Christian Teitzel. Images from DVD *Touching Soap Films*, Springer, 1995.

p. 238: AF archive / Alamy

Index